Sustainable Design

Tomayess Issa • Pedro Isaias

Sustainable Design

HCI, Usability and Environmental Concerns

Tomayess Issa
Curtin University
School of Information Systems
Perth, Australia

Pedro Isaias
Universidade Aberta - Portuguese Open
  University
Department of Social Sciences
  and Management
Lisbon, Portugal

Additional material to this book can be downloaded from http://extras.springer.com.

ISBN 978-1-4471-6992-5   ISBN 978-1-4471-6753-2   (eBook)
DOI 10.1007/978-1-4471-6753-2

Springer London Heidelberg New York Dordrecht

Springer-Verlag London Ltd. is part of Springer Science+Business Media (www.springer.com)

Preface

The quest for sustainable and intuitive interaction between humans and the technologies they habitually engage with vibrates throughout the plethora of publications currently available in the field of HCI research and development. The first flurry of research focusing on sustainability by the HCI community began around 2007, and by 2009, using the ACM guidelines, Goodman declared at least 120 papers to be related to sustainable HCI. Since then, this specialised field in HCI has flourished. Whilst such an intensity of interest might portend the development of numerous applications and possible solutions, the lack of appropriate support and guidance in the form of usable models for implementation in the business sector has hindered any real progression.

In an age where technologies are designed with a clean minimalist design edge, the reality is that the components used to develop ICT technologies such as phones, tablets, computers, etc. pose substantial risk to the health of our society and the environment in which we live. E-waste is quickly becoming our foremost solid-waste entity. Such issues are at the very nexus of the concerns expressed in this text. As a discipline, HCI focuses on the ways in which humans interact with technologies, and many textbooks have successfully engaged with the central precepts in the design, development, and usability of interactions. However, whilst perhaps devoting a few pages, maybe even a chapter, few HCI texts have hitherto considered sustainability and resource management essential to their remit. This text changes all that.

In amongst this surfeit of available texts, this book stands out as a useful 'handbook' of information that is both explanatory and relevant to contemporary applications of HCI.

I am delighted to introduce the reader to this text. Here, in easily accessible language, the authors have offered the reader entry to a sometimes difficult conceptual world. Students in particular will find it possible to extend their research beyond the confines of the undergraduate classroom and the routine HCI textbook. However, this book will also assist those requiring a rather more rigorous and complex resource and those at the very interface of human-computer interaction and sustainable use, such as industry and design practitioners. In addition, academics seeking critical evaluation of the potential and actual ramifications of our design and its impact on our use of technologies will encounter some useful methods for reflecting on their own research practice.

The authors consider not only the design and development issues routinely discussed in HCI texts but also propose a series of methodologies to assist the reader in developing applications that adhere to sustainable guidelines. Not content to merely offer 'ideal' solutions often impossible to implement, the authors proffer models designed for usability. Students in particular will benefit from the authors' definition of the phases and activities required and, further, the most appropriate tools and techniques for development of sustainable interfaces.

In doing so, the authors discuss the ways in which technologies can still meet the needs of our society yet ensure that in the process, natural resources are neither damaged nor depleted. Consideration is also given to source reduction through reducing the wastage in the production and consumption of technology via exemplary HCI design.

This is the text we have been waiting for.

This is a text for our future.

Norway   Katherine Blashki

Acknowledgements

Tomayess Issa owes special gratitude first to the Almighty and her parents: V. Rev. Fr. Boutros Touma Issa and Bathqyomo Marine Khoury-Issa; her sister Dr. Theodora, her brother Dr. Touma, his wife Siba, and their daughter Talitha; her Sister Tamara, her husband Tony, and their children Tabitha, Antoinette, Jacob, and finally to her brother Theodore (FCPA), his wife Mary, and their children Cephas and Mary for their continuous support and encouragement since without their help, this work would never have been completed. Tomayess would like to acknowledge the support of the school of Information Systems – Curtin University, Australia, for their inspiration and encouragement.

Pedro Isaias would like to acknowledge Universidade Aberta and Curtin University for their support to this author specifically. Also a special acknowledgement to Sara Pifano from ISRLab for her continuous support. Finally, he would like to acknowledge the support of his family. Without their assistance, this work would not be possible at all.

Perth, Australia   Tomayass Issa
Lisbon, Portugal   Pedro Isaias

Contents

Chapter 1
Introduction

Abstract This book will examine the importance of Human Computer Interaction, Usability, and Sustainability, including sustainable design, in the Information Communication and Technology sector (ICT). ICT usage by businesses and individuals has become a significant instrument for searching, conducting research, communication, entertainment, commerce and information. The recycling of ICT usage is becoming a major dilemma for businesses and individuals, since it is not simply a matter of concern for environmental damage or a solution to an environmental problem. Designers, businesses, and individuals must collaborate in making a concerted effort to tackle the environmental concerns by developing new ICT technologies with sustainable design in their agenda to meet the needs of businesses and individuals both currently and in future. This book discusses sustainable design features as well as the New Participative Methodology for Sustainable Design.

1.1 Introduction

Computer technology, internet technology, and systems are essential tools in the twenty-first century since businesses and individuals have come to depend increasingly on these technologies compared with the traditional systems used to achieve the same ends. The current technology is more capable of managing and assisting businesses and individuals to complete their tasks far more efficiently. Not only is there a proliferation of stand-alone computers; networking on a global scale has increased enormously as a result of the Internet, World Wide Web, social networks, mobile systems, Intelligent Environments and others. The increase of ICT usage throughout the world has presented a new challenge to HCI researchers and practitioners to match businesses and individual needs and ensure that the new ICT technologies are more sustainable for both current and future needs. HCI is the study of the interaction between humans and complex technology in order to examine how the current input and output of technologies influence the interaction between users and interface. HCI draws on many disciplines but it is in "computer science and systems design that it must be accepted as a central concern, and HCI involves the design, implementation and evaluation of interactive systems in the context of the user's task and work" (Dix et al. 1993, p. 4).

© Springer-Verlag London 2015
T. Issa, P. Isaias, *Sustainable Design*, DOI 10.1007/978-1-4471-6753-2_1

Therefore, HCI researchers should consider within their discipline not only productivity and customer satisfaction, but also human factors that affect "acquisition, disposal, renewal, and re-use and design for sustainability" (Dillahunt et al. 2010, p. 1). In addition, they should assist to create and develop technologies which are more effective and efficient and should study the "social and communal aspects of technology use and effective and aesthetic aspects of design" (Sengers et al. 2006, p. 1683). To achieve this, they must consider the different perspectives of users and designers in order to understand their notions of design, attitudes, ethnography, user empathy, and seek to develop new technologies that address sustainability goals for the current and future generations (Busse et al. 2009; Sengers et al. 2009a).

Hence, HCI researchers, businesses, and individuals should add to their notions of design the concept of "green" technologies, since the current technologies are adversely affecting and causing major problems to the environment. In addition, sustainability principles should be applied to the system design to ensure that the new design is more sustainable, user friendly, safe, efficient, effective, and usable for businesses and individuals. This is done by studying and understanding potential users' desires and requirements. Furthermore, this book will examine the importance of HCI, Usability, and Sustainability in respect to design systems, thereby raising the awareness of HCI practitioners and academics regarding the development of new technologies, bearing in mind the future generations. In addition, a new sustainable design model will be developed to promote the notions of HCI, Usability, and Sustainability when developing new devices now and in future.

This book is organized as follows: Introduction, HCI, Usability, User Participation in the System Development Process, Physical, Cognitive Affective Engineering, Color, Prototyping and Navigation, Guidelines and Principles Design, Evaluation and Testing; Task Analysis, Models, and Methodologies and the New Participative Methodology for Marketing Websites (NPMMW), the New Participative Methodology for Sustainable Design (NPMSD) and Future ICT.

1.2 Human-Computer Interaction

Human-Computer Interaction was adopted in the mid-1980s as a means of describing this new field of study. HCI "is a discipline concerned with the design, evaluation and implementation of interactive computing systems for human use and with the study of major phenomena surrounding them" (Preece et al. 1994, p. 7). However, this field is now "concerned with understanding, designing for, and evaluating a wider range of user experience aspects" (Sharp et al. 2011, p. 18). Therefore, the reason for studying HCI in the development process is to create interactive computer systems that are usable as well as practical (Head 1999).

The term 'HCI' relates to several stages in the development process, including the design, implementation and evaluation of interactive systems, in the "context of the user's task and work" (Dix et al. 2004, p. 4). According to Vora (1998), HCI implementation requires a massive range of skills, including an understanding of the potential users, their tasks, and environments, software engineering capabilities, and graphical interface.

Designers often have a poor understanding of HCI issues; therefore, designers need to know how to think in terms of future users' needs, values and supportable tasks and how to translate that knowledge into an executable system. This can be accomplished by establishing a good interface design to let the user interact and deal with the user interfaces without any difficulties and to give the user more control of the site.

The main purpose of using HCI in the design is to develop an efficient and effective user interface to suit the needs and requirements of the users. To achieve these features, HCI specialists need to involve the users in their design, integrating different kinds of knowledge and expertise, and making the design process iterative (Preece et al. 1994). It was noted that HCI design should be user-centered to integrate knowledge from different disciplines and be highly iterative, and include an effective usability evaluation. This type of process will allow for feedback regarding negative and positive aspects of prototypes. It is important that the way in which people interact with computers be intuitive and clear. However, the designing of appropriate HCI is not always straightforward, as the many poorly designed computer systems testify. One of the challenges of HCI design is to keep abreast of technological developments and to ensure that these are harnessed for maximum human benefit.

The goals of HCI are to produce usable, safe and functional systems. These goals can be summarized as safety, utility, effectiveness, efficiency and appeal. These goals focus on the services that the system provides and how quickly the tasks can be achieved, and ensuring that users like the system. By the same token, Haklay (2010, p. 5) indicated that HCI aims to create systems which provide functionality to meet the needs of businesses and individuals. In addition, in order to develop or improve their design, HCI specialists should understand how system design can support users in an effective and efficient manner, and how users intend to use computers systems. Finally, Bodker, Byrne and Boye (cited in (Maceli and Atwood 2011) describe the three waves of HCI: humans as factors, actors and crafters. Therefore, all information interfaces including websites should have a good interaction with users and vice-versa to effectively ensure efficiency and safety, and make them more enjoyable for users.

1.3 Usability

Usability refers to the "quality of the interaction in terms of parameters such as time taken to perform tasks, number of errors made and the time to become a competent user" (Benyon et al. 2005, p. 52). Alternatively, Usability "is a quality attribute that assesses how easy user interfaces are to use. The word 'usability' also refers to methods for improving ease-of-use during the design process" (Nielsen 2003).

Furthermore, Shackel (2009, p. 340) indicates that usability is the "capability in human functional terms to be used easily and effectively by the specified range of users, given specified training and user support, to fulfill the specified range of tasks, within the specified range of environmental scenarios".

The usability evaluation stage is an effective method by which a software development team can establish the positive and negative aspects of its prototype releases, and make the required changes before the system is delivered to the target users. Usability evaluation is about observing users to "see what can be improved, what new products can be developed" (McGovern 2003). It is "based on human psychology and user research" (Rhodes 2000). HCI specialists "observe and talk with participants as they try to accomplish true-to-life tasks on a site, and this allows them to form a detailed picture of the site as experienced by the user" (Carroll 2004).

From the user's perspective, usability is a very important aspect of the development process as it can mean the difference between "performing a task accurately and completely or not" and the user "enjoying the process or being frustrated"(Usability First 2002). Alternatively, if usability is not an integral part of user interface design, then users will become very frustrated working with it. In general, usability is an essential concept in HCI and is concerned with making systems easy to learn, easy to use, and with minimal error frequency and severity. In order to develop a successful system with good usability, HCI specialists need to understand and realize various factors, namely organizational, social and psychological factors that determine the extent to which people effectively operate and make use of computer technology. They need to develop tools and techniques to help designers ensure that computer systems are suitable for the activities for which people will use them, and achieve efficient, effective and safe interaction in terms of both individual Human Computer Interaction and group interaction. These factors should be considered very carefully at the design stage, as most of the users should not have to change radically to 'fit in' with the system; rather, the system should be designed to meet their requirements (Preece et al. 1994).

Furthermore, Sharp et al. (2011) indicate that usability goals should be considered by designers and HCI specialists to ensure that the user interface is easy to learn and remember, effective and efficient to use, and with fewer errors and good utility. These goals can be applied to the design of an interactive system in order to promote its usability. Therefore, these principles are intended to give more assistance and knowledge to system developers regarding the system design. Together with the above principles, an important additional key factor is Utility. Utility refers to the functionality so users can "do what they need or want to do" (Preece et al. 2002, p. 16). In other words, "does it do what users need?" (Nielsen 2003). Hence, usability and utility are equally important in the development process and need to be integrated.

Finally, it was noted that HCI and Usability are essential factors to consider when designing and developing a user interface, which is more efficient and effective and produces user satisfaction rather than frustration. In order for the interface to have these attributes, the potential users should participate in the design from the outset. Folstad et al. (2010) and Issa et al. (2010) reiterate that user participation is essential in the system development process and users should be present during this process to share their opinions, especially from the initial planning stage through to the maintenance stages and procedures.

Furthermore, according to Issa et al. (2010), user participation in the system development process will prevent user frustration, thereby reducing training time, and ensuring that the system is designed to match users' requirements. Finally, Nies and Pelayo (2010) posit the same notion that it is necessary to involve users in the system development process so that the design meets their requirements.

1.4 Sustainability

Before discussing the term 'sustainable design', firstly we need to discuss the notion of 'sustainability', since these two concepts are related in terms of benefitting human and natural resources that will be needed in the future (Weybrecht 2010). Gro Harlem Brundtland from the World Commission on Environment and Development first coined the term 'sustainability' in 1983. Brundtland's report urged businesses and individuals to progress toward economic development in a way that could be sustained without destroying the natural resources or the environment for the next generation.

Erek et al. (2009, p. 2) define sustainability as "a survival assurance meaning that an economical, ecological or social system should be preserved for future generations and, thus, necessary resources should only be exploited to a degree where it is possible to restore them within a regeneration cycle". This suggests that businesses and individuals must protect the current infrastructure so that it can be re-used by the next generation. The notion of sustainability is highly significant in the twenty-first century since, increasingly, businesses and individuals are now required to think in terms of delivering "solutions rather than products, and seek to define their markets in terms of customer activities and outcomes rather than products and services" (Jeffers 2009, p. 263).

The integration of sustainability in businesses and in individuals' strategies will be highly advantageous in terms of cost reduction, resources preservation, conformity to legislation, improvement of reputation, maintaining happier customers and stakeholders, attracting capital investment and capitalizing on new opportunities (Nidumolu et al. 2009; Sharma et al. 2010; Smith and Sharicz 2011). Finally, Kendall and Kendall (2010) indicated that sustainability will assist businesses, stakeholders, individuals and society in general.

The integration and application of sustainability strategy in business should suit project needs and business proposals of a particular division or even the whole company. According to Weybrecht (2010), the adoption and application of sustainability in businesses will achieve the following advantages: cost reduction; preservation/saving of resources; compliance with legislation; enhanced reputation that differentiates businesses; securing quality employees; satisfy customer needs;

meeting of Stakeholder expectations; attracting of capital investment; and capitalizing on new opportunities. These advantages will make the business unique in the market locally and globally, since sustainability is already, a part of how business is done; the nature of the business is not as important as its ability to continue. Currently, the potential high cost of sustainability for both the business and society since multiple benefits will be achieved by integrating sustainability in the business strategy. However, sustainability will be very strong when it is embedded into the strategy and culture of a business with the full cooperation of the CEO.

To integrate the sustainability factor in the business strategy, the project manager should collect all the necessary information about what is happening in his/her company at all levels of the business hierarchy. Once the required information has been collected, it is necessary to secure everyone's cooperation so that all employees and management have the same positive attitude toward sustainability. The project manager must pick the correct moment to disseminate the notion of sustainability adoption throughout the organization. The advantages and disadvantages of integrating sustainability in a business strategy should be put to management whose role it is to inform staff of any changes that this requires. Furthermore, in terms of sound business practice, the different attitudes of staff together with their roles, backgrounds, and personalities should be taken into account.

Moreover, the project manager must make a strong case by outlining the benefits of a sustainability policy, and the disadvantages if the organization does not address this issue. This adoption of sustainability in the organization structure will be useful when hiring new staff.

Finally, sustainability as an integral part of an organization's strategy requires understanding, consideration, and tolerance at all levels of the organization as well from its stakeholders. The strategy should be easy, straightforward, dynamic, and easy to implement. Finally, patience must be exercised when changing the mindset and attitudes of staff and stakeholders in terms of introducing sustainability strategies.

According to Moscardo et al. (2013), sustainability requires a long-term orientation and commitment to changing the way businesses conduct their activities in order to balance the needs of the current personnel with those of future generations. Furthermore, there should be recognition that business is part of a complex system in which environmental, social, and economic activities are common. Part of the strategy awareness and training should be available to improve knowledge about sustainability. This learning should not be limited to staff; specific training should be available to stakeholders and the community to make them aware of all the issues concerning sustainability, since the needs of the business should and must match the needs of stakeholders, society, the economy and the environment. Implementing sustainability in business strategy will enhance natural capital and improve a company's reputation in the market nationally and internationally.

Finally, sustainability is a complex area that is continually changing and growing. This means each person in an organization should participate in this change from training, learning, considering the benefits and risks, green washing and changing the mindset.

1.5 Sustainable Design

According to Nathan et al. (2008), the terms "sustainable" and "green" are used ubiquitously within businesses and by individuals locally and globally. Currently, these notions play a major role in businesses and individual strategies; therefore any design should ensure that whatever is created and developed should first meet the current users' and businesses' requirements and, of course, those of the next generation.

Stelzer (2006, p. 4) defines sustainable design as the "fundamentally a subset of good design. The description of good design will eventually include criteria for the creation of a healthy environment and energy efficiency." Silberman and Tomlinson, (2010, p. 3470) discuss and argue the relationship between sustainable design and HCI, confirming that previously HCI researchers were concerned with "What do users do? When? How often? Why? How do they feel about it? What do know about what they are doing? How do they know?"

Nowadays, however, HCI researchers should understand the relationship between users and technologies, and how this can assist designers to simplify more sustainable user practices. Moreover, DiSalvo et al. (2010) confirm that HCI researchers and top management should be encouraged to collaborate in the design and development of applications, interfaces, equipment and services with more sustainable effects; in addition, this design should comply with the principles of economic, social and ecological sustainability. Smith and Sharicz (2011) posit that HCI and Information Technology researchers and professionals must take into consideration the environmental impact of the design of current and future technologies, so that practitioners are aware of the environmental impact of the technologies they use. Most importantly, sustainable design should meet users' needs. Sustainable design needs awareness and innovation among designers and users. Awareness can produce opportunities to be unique and exceptional in design, and this can lead to creativity and innovations in research. Awareness of the need for change can contribute to the improvement of the environment, to social equity and to growth and profit in the expanding global community. This awareness will lead the designers to action orientation, learning, and excitement, and to a new level of caring based upon new knowledge and commitment. To achieve the above, participation in sustainable design is essential, and designers must take into account the opinions and perspectives of potential users to assist with the design, since designers cannot act by themselves.

Currently, the world is under pressure from human actions that threaten sustainability. At the global level, the quantity of e-waste generation in 2014 was around 41.8 million tones, and 4 billion people were covered by national e-waste legislation. This number will be increased to 49.8 by 2018, meaning an increase growth rate of 4–5 % if developers still maintain the status quo when designing, without integrating sustainability in their practice agenda (Baldé et al. 2015). Figure 1.1 shows the total e-waste per category in 2014. Small devices such as USB-sticks, phones, and electronic toothbrushes have the highest rate compared to 1.0 MT for the lamps.

Fig. 1.1 E-waste per category in 2014 (Adopted by Baldé et al. (2015). Prepared by the authors)

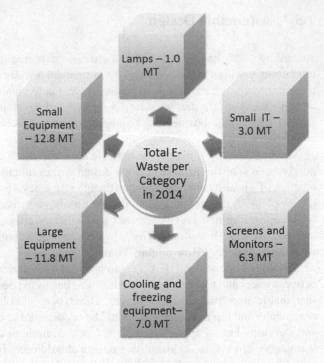

Fig. 1.2 E-waste generation per continent in 2014 (Adopted by Baldé et al. (2015). Prepared by the authors)

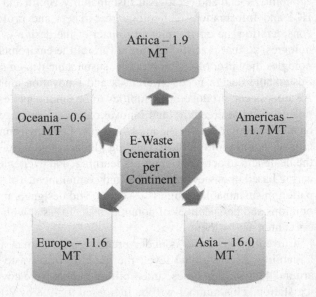

Furthermore, Fig. 1.2 shows the e-waste generated per continent; Asia generated 16 MT in 2014, while Oceania generated only 0.6 MT in 2014.

As shown by the results presented in Figs 1.1 and 1.2, the world is experiencing a great many transformations as a result of human unsustainable actions; therefore, a plan of action should be implemented to change the way we live. Therefore designers, users, and organizations, should focus their minds and commitment on designing objects and devices that comply with the principles of social, economic and ecological sustainability.

Finally, sustainable design will be the way to make our world better. However, in order to achieve this, we need to have the right motivation, awareness, knowledge, commitment, trust, and loyalty. People need to act quickly to think about good and sustainable design by adopting sustainability in their business strategy in order to conserve raw materials for the next generation.

1.6   Methodology

For this book, an online survey is employed to examine users' attitudes toward sustainability and sustainable user interface design in Australia. An online survey will assist the authors to identify the new factors, which are required for the new sustainable design model. The online survey has been created based on the findings of the literature; and is divided to three sections; background; sustainable design, and advantages and disadvantages of sustainability. Employing an online survey in this study allows the users to identify the new factors for the new sustainable model and identify the new theoretical significance of this book. The online survey can offer greater anonymity, is less expensive, and is more accessible (O'Brien and Toms 2010; Kocher 2015; Issa 2013). However, technical failure, computer viruses, internet crimes, hacking, and privacy are considered the disadvantages of online surveys, and these factors can reduce the response rate (Fan and Yan 2010).

1.7   The Initial Sustainable Step in the New Participative Methodology for Sustainable Design

Sustainability is now generally accepted by most organizations as an important part of corporate citizenship. The concept of sustainability is based on the notion that our actions should not cause irreparable harm to our social and environmental infrastructure. It calls for our responsibility and action to improve or change our current way of living to avert social, environmental, and ecological crises. The term 'sustainable development' was first referred to in 1987 in the Brundtland Report on 'Our Common Future', where it states that 'sustainable development is development that meets the needs of the present without compromising the ability of future generations to meet their own needs'. Incorporating sustainable strategy with emerging technologies is becoming the norm in contemporary businesses (Newton 2003).

To perform this effectively, and deriving from Dyllick and Hockerts (2002) and McDonough and Braungart (2002) models of corporate sustainability, Young and Tilley (2006) proposed an integrated model of corporate sustainability which links together six criteria that a sustainable business will need to satisfy. The six criteria are (1) eco-efficiency, (2) socio-efficiency, (3) eco-effectiveness, (4) socio-effectiveness, (5) sufficiency and (6) ecological equity. However, further theoretical development is still under way in order to create an effective, integrated approach to applying the six criteria. Erek et al. (2009, p. 2) stated, "Sustainability has been extensively discussed within corporate management under the synonyms of corporate social responsibility (CSR), greening the business eco-efficiency or eco-advantage." To ensure that organizations develop and adhere to a sustainable development strategy, management should consider aspects of value creation that would benefit its employees, users and stakeholders by encouraging all participants to be environmentally and socially responsible corporate citizens.

In line with the integration of a sustainability strategy into technology, various studies from Human Computer Interaction, Usability and Sustainability were examined and investigated to study the ICT impacts on environment (Ramani 2010; Bevan 2001; Bodker 2006; Dillahunt et al. 2010; DiSalvo et al. 2010; Mann 2009; Nathan et al. 2008; Sengers et al. 2006, 2009b; Silberman and Tomlinson 2010; Wilson and Borras 1998; Dix et al. 1993; Gerlach and Kuo 1991; Te'eni et al. 2007). It was noted that the recycling of ICT usage is becoming a major dilemma for businesses and individuals, since it is not simply a matter of concern for environmental damage or a solution to an environmental problem.

Designers, businesses, and individuals must collaborate in making a concerted effort to tackle the environmental concerns by developing new ICT technologies with sustainable design in their agenda to meet the needs of businesses and individuals both currently and in future. Therefore, this book will discuss and present a New Participative Methodology for Sustainable Design for smart new technology and portable devices. From a review of the current literature (Gauthier 2015; Kemp 2015; Pan et al. 2015; Shaw et al. 2015; Stapledon et al. 2015; Wang et al. 2015a, b; Stelzer 2006; Nidumolu et al. 2009; Issa 2014; Issa and Isaias 2014; Comm and Mathaisel 2015; Wals 2014) the initial factors for the sustainable step have identified from design, safety, manufacture and energy, recycle efficiency and social (see Fig. 1.3).

These critical factors will assist to develop the first draft of the New Participative Methodology for Sustainable Design The authors will add the new characteristics and critical factors, which belong to the new sustainable model under the design stage under the new Participative Methodology for Marketing Websites' (NPMMW) – (See Fig. 1.4). NPMMW methodology includes all the necessary stages and steps, which are required to develop an efficient and effective device.

Figure 1.5 illustrates the first draft of the new Sustainable Model, which will be part of the design stage under the NPMMW methodology. This model will use all the stages and steps, which belong to the NPMMW model to ensure that the new devices meet users' requirements and needs.

According to Stelzer (2006), Sustainability is primarily a subset of design. Design is an exercise in meeting the challenges inherent in any situation that requires improvement or mediation. Ultimately, any design solution will need to create products and environments for a living earth with limited resources. The criteria for

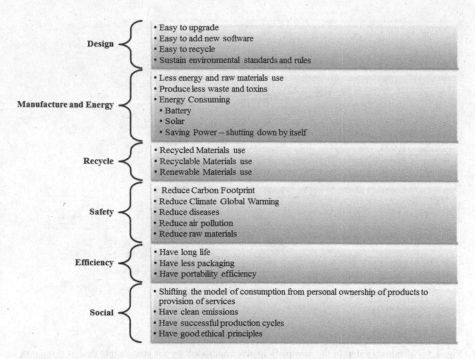

Fig. 1.3 Initial factors for the sustainable step (Prepared by Tomayess Issa)

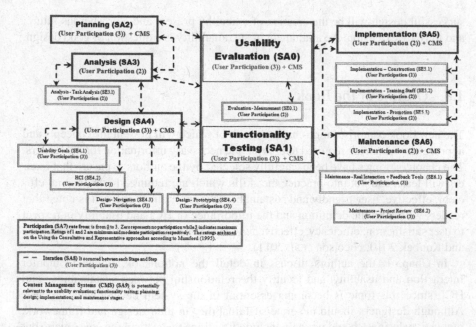

Fig. 1.4 The New Participative Methodology for Marketing Websites' (NPMMW) (Prepared by Tomayess Issa)

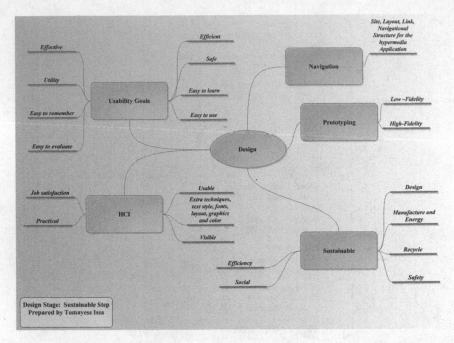

Fig. 1.5 Sustainable step in the new participative methodology for sustainable design (Prepared by Tomayess Issa)

successful design will be the creation of a healthy present and a prosperous future; and thus, by extension, the attainment of sustainability is a question of good design.

1.8   Outline of the Book

The new book comprises nine chapters, each of which will present the concepts and approaches, which are required to provide the necessary information for the readers. The chapters' topics have been carefully selected by the authors to ensure that readers will learn and put into practice the skills which are required to develop an efficient, effective, user friendly and sustainable design. From Fig. 1.6, it is noted that readers will learn the definition and the importance of HCI and usability in respect to user satisfaction, efficiency, effectiveness and user friendliness of the system (Lee and Koubek 2010; Nicolson et al. 2011).

In Chap. 1, the authors discuss in detail the notions of Human Computer Interaction and usability, and identify the relationship between sustainability and HCI, since this topic is becoming essential in the system development process. Although designers should integrate sustainability in their design and framework, innovative designs should not only include functions that satisfy the consumers, but should also be sustainable (Ramani 2010). Therefore, designers, users and top

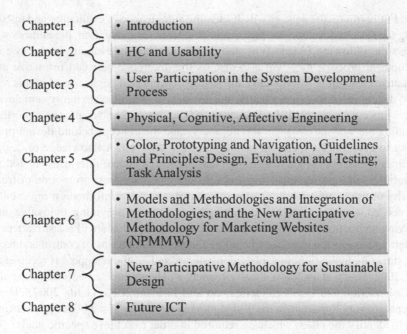

Chapter 1	• Introduction
Chapter 2	• HC and Usability
Chapter 3	• User Participation in the System Development Process
Chapter 4	• Physical, Cognitive, Affective Engineering
Chapter 5	• Color, Prototyping and Navigation, Guidelines and Principles Design, Evaluation and Testing; Task Analysis
Chapter 6	• Models and Methodologies and Integration of Methodologies; and the New Participative Methodology for Marketing Websites (NPMMW)
Chapter 7	• New Participative Methodology for Sustainable Design
Chapter 8	• Future ICT

Fig. 1.6 Outline of the book – table of contents

management must, together, work smarter and harder and more creatively if "we are going to help save our planet from ourselves" (Ramani 2010, p. 1).

Several scholars (Dillahunt et al. 2010; Huh and Ackerman 2009; Sharma et al. 2010; Thatchenkery et al. 2010), propose that an agreement should be developed between designers, users and top management encouraging them to work collaboratively on a sustainable interface, application and equipment to meet the current and future generation in order to minimize damage done to our planet.

Chapter 2 encourages readers to learn the principles and guidelines for Human Computer Interaction and Usability in the system development process. Chapter 3 focuses on user participation in the system development process by obtaining opinions about and attitudes to the design in order to prevent potential user frustration.

Chapter 4 examines the differences between physical, cognitive and affective engineering, since these topics will assist readers to understand that design is not limited to layout, navigation and colour, but that other aspects should also be taken into consideration in the design process. These topics will engender discussion about the interaction and relationship between human and machine, ergonomics, and development concerns such as memory, attention span of users, and reduction of complexity between the goals of cognitive engineering, speed and accuracy and finally, effectiveness, i.e. making the interface more attractive, beautiful, entertaining, enjoyable, engaging and fun (Te'eni et al. 2007).

Chapter 5 discusses the importance of colour, navigation and prototyping in the system development process (Bonnardel et al. 2011; Cyr et al. 2010), as designers and users should be satisfied with the final sketches before coding and implementation occur. Furthermore, the authors will discuss the significance of evaluating and testing during the system development process. In respect to the evaluation, the authors will address the following issues: Why and what and when to evaluate in the system development process; they will also discuss the difference between formative and summative evaluation. Additionally, the testing concept will be discussed in this section to distinguish between evaluation and testing and their place in the system development process (Issa et al. 2010; Petre et al. 2006). To assess and evaluate an interface (including the website), readers should understand the concept behind design principles and guidelines which will be introduced in this unit. A knowledge of design principles is essential since readers will learn how to evaluate interfaces (including the websites) in a professional way from different perspectives: promotion of trust, diversity of users, affordability and performance, matching information representations needed with that presented, designing for errors, and providing, enjoyable, and satisfying interaction. On the other hand, the design guidelines will assist readers to evaluate and assess the interface (including the website) in terms of control and feedback, direct manipulation, metaphor, consistency and aesthetic appeal (Preece et al. 1994, 2002; Te'eni et al. 2007). Finally, readers will learn three aspects of task analysis: Task, Action and Goals (Shneiderman and Plaisant 2010; Galitz 2007). These concepts are very important in the design process since they assist both designers and users to identify the tasks, which are required in order to achieve specific goals.

To ensure that interfaces are developed successfully without causing frustration to users, Chap. 6 introduce a series of methodologies to demonstrate the stages and steps, which are required to develop a system in a sequential manner, by defining the activities, method and techniques, and tools which are required to develop these interfaces.

Chapter 7 discusses the New Participative Methodology for Sustainable Design and identify the new factors, which are required to develop a sustainable design now and in future.

Furthermore, the authors will continue to introduce other topics to the new unit program, i.e. social and global issues and social networking including Web 2.0 and 3.0 in Chap. 8. The former topic will include the following social aspects of information systems and how HCI can ameliorate these aspects: anxiety, alienation, potency and impotency of the individual, complexity and speed, organizational and societal dependence, valuing human diversity, privacy, accessibility, accountability and property, and the social and global impacts of the Internet (Te'eni et al. 2007; Thakurta 2010). While the latter topic will be concerned with how social, networking (Web 2) is becoming a critical strategy in teaching, especially since these tools can assist in teaching and learning, not just in social life. Furthermore, Web 3.0 will be introduced to readers since this new technology is more creative and dynamic compared with Web 2.0 (Kearns and Frey 2010; Rego et al. 2010).

Finally, this book concentrates on establishing and consolidating the relationship between HCI, Usability and Sustainable design, and sharing the latest information in respect to the previous topics, since the majority of HCI authors are keen to develop frameworks, tools, techniques, and models to meet the sustainable design requirements.

1.9 Conclusion

This chapter discussed and examined the concepts, which are required for sustainable design. To identify the new sustainable model, an initial model is discussed and an online survey is distributed in Australia to examine users' attitudes to sustainability and sustainable user interface design. The online survey results will be discussed in Chap. 8; and later we identify the new factors, which are required for new sustainable model. Finally, this chapter presented an overview of this book.

References

Baldé CP, Wang F, Kuehr R, Huisman J (2015) The global e-waste monitor. United Nations University IAS-SCYCLE, Bonn

Benyon D, Turner P, Turner S (2005) Designing interactive systems. Pearson Education Limited, Edinburgh

Bevan N (2001) International standards for HCI and usability. Int J Hum Comput Stud 55(4):533–552. doi:10.1006/ijhc.2001.0483

Bodker S (2006) When second wave HCI meets third wave challenges. In: Proceedings of the 4th nordic conference on human computer interaction: changing roles, Oslo

Bonnardel N, Piolat A, Le Bigot L (2011) The impact of colour on Website appeal and users' cognitive processes. Displays 32(2):69–80. doi:10.1016/j.displa.2010.12.002

Busse D, Blevis E, Howard C, Dalal B, Fore D, Lee L (2009) Designing for a sustainable future. In: Proceeding of the seventh ACM conference on creativity and cognition, Berkeley, pp 493–494

Carroll M (2004) Usability testing leads to better ROI. http://www.theusabilitycompany.com/news/media_coverage/pdfs/2003/NewMediaAge_270303.pdf. Accessed 1 Sept 2014

Comm CL, Mathaisel DF (2015) Designing an engineering system for sustainability. Appl Mech Mater 704:474–478, Trans Tech Publ

Cyr D, Head M, Larios H (2010) Colour appeal in website design within and across cultures: a multi-method evaluation. Int J Hum Comput Stud 68:1–21

Dillahunt T, Mankoff J, Forlizzi J (2010) A proposed framework for assessing environmental sustainability in the HCI community. In: CHI 2010, Atlanta, pp 1–3

DiSalvo C, Sengers P, Hronn Brynjarsdottir P (2010) Mapping the landscape of sustainable HCI. In: CHI 2010, Atlanta, pp 1975–1984

Dix A, Finlay J, Abowd G, Beale R (1993) Human computer interaction. Pearson Prentice Hall, Harlow

Dix A, Finlay J, Abowd G, Beale R (2004) Human-computer interaction, 3rd edn. Pearson Education Limited, Harlow

Dyllick T, Hockerts K (2002) Beyond the business case for corporate sustainability. Bus Strateg Environ 11(2):130–141

Erek K, Schmidt N-H, Zarnekow R, Kolbe LM (2009) Sustainability in information systems: assortment of current practices in IS organizations. In: Proceedings of the Americas conference on information systems (AMCIS), San Francisco, pp 1–9

Fan W, Yan Z (2010) Factors affecting response rates of the web survey: a systematic review. Comput Hum Behav 26(2):132–139

Folstad A, Anda B, Sjoberg D (2010) The usability inspection performance of work-domain experts: an empirical study. Interact Comput 22(2):75–87

Galitz W (2007) The essential guide to user interface design: an introduction to GUI design principles and techniques. Wiley, New York

Gauthier G (2015) A usability evaluation of a website focusing on the three initial steps of the conflict/resolution process for union members. http://scholarspace.manoa.hawaii.edu/bitstream/handle/10125/35855/Gauthier_Final_Paper_Scholarspace.pdf?. Accessed 1 Sept 2014

Gerlach JH, Kuo F-Y (1991) Understanding human computer interaction for information systems design. MIS Q 14(4):526–549

Haklay M (2010) Interaction with geospatial technologies. Wiley, Chichester/Hoboken

Head AJ (1999) Design wise. Thomas H Hogan Sr, Medford

Huh J, Ackerman M (2009) Challenges in sustainability: understanding users' appropriation and maintenance work of computational artifacts. In: CHI 2009. pp 1–3

Issa T (2013) Online survey: best practice. In: Information systems research and exploring social artifacts: approaches and methodologies, IGI Global, pp 1–19. doi:10.4018/978-1-4666-2491-7.ch001

Issa T (2014) Onine shopping and human factors e-commerce platform acceptance : suppliers, retailers, and consumers. Springer, UK

Issa T, Isaias P (2014) Promoting human-computer interaction and usability guidelines and principles through reflective journal assessment, Emerging research and trends in interactivity and the human-computer interface. IGI Global, Hershey, pp 375–394. doi:10.4018/978-1-4666-4623-0.ch019

Issa T, Turk A, West M (2010) Development and evaluation of a methodology for developing marketing websites. In: Martako D, Kouroupetroglou G, Papadopoulou P (eds) Integrating usability engineering for designing the web experience: methodologies and principles. IGI Global Publishing, Hershey, pp 103–123

Jeffers P (2009) Embracing sustainability – information technology and the strategic leveraging of operations in third-party logistics. Int J Oper Prod Manag 30(3):260–287

Kearns L, Frey B (2010) Web 2.0 technologies and back channel communication in an online learning community. Tech Trends 54(3):41–54

Kemp S (2015) Digital, social and mobile worldwide in 2015. http://wearesocial.net/tag/sdmw/. Accessed 22 Jan 2015

Kendall K, Kendall J (2010) Forms of government and systemic sustainability: a positive design approach to the design of information systems. Adv Appreciative Inq 3:137–155

Kocher M (2015) Recipes and research: a survey of cookbook collection users. J Agric Food Inf 16(1):53–59

Lee S, Koubek RJ (2010) The effects of usability and web design attributes on user preference for e-commerce web sites. Comput Ind 61(4):329–341

Maceli M, Atwood M (2011) From human factors to human actors to human crafters. Paper presented at the iConference, Seattle

Mann S (2009) SIGCHI workshop position paper: sustainable practitioners in HCI. Paper presented at the CHI 2009, Boston

McDonough W, Braungart M (2002) Design for the triple top line: new tools for sustainable commerce. Corp Environ Strateg 9(3):251–258

McGovern G (2003) Usability is good management. http://www.gerrymcgovern.com/nt/2003/nt_2003_04_07_usability.htm. Accessed 1 May 2015

Moscardo G, Lamberton G, Wells G, Fallon W, Lawn P, Rowe A, Humphrey J, Wiesner R, Pettitt B, Don C, Renouf M, Kersham W (2013) Sustainability in Australian business: principles and practice. Wiley, Milton

Nathan L, Belvis E, Friedman B, Hasbrouck J, Sengers P (2008) Beyond the hype: sustainability and HCI. In: CHI 2008, Italy, pp 1–4

Newton LH (2003) Ethics and sustainability, sustainable development and the moral life. Basic Ethics in Action. Prentice-Hall, Inc., New Jersey

Nicolson D, Knapp P, Gardner P, Raynor D (2011) Combining concurrent and sequential methods to examine the usability and readability of websites with information about medicines. J Mixed Methods Res 51(1):25–51

Nidumolu R, Prahalad CK, Rangaswami MR (2009) Why sustainability is now the key driver of innovation. Harv Bus Rev 87(9):57–64

Nielsen J (2003) Usability 101. http://www.useit.com/alertbox/20030825.html. Accessed 1 May 2015

Nies J, Pelayo S (2010) From users involvement to users' needs understanding: a case study. Int J Med Inform 79(76–82)

O'Brien H, Toms E (2010) The development and evaluation of a survey to measure user engagement. J Am Soc Inf Sci Technol 61(1):50–69

Pan Y, Xu Y, Wang X, Zhang C, Ling H, Lin J (2015) Integrating social networking support for dyadic knowledge exchange: a study in a virtual community of practice. Inf Manag 52(1):61–70. doi:http://dx.doi.org/10.1016/j.im.2014.10.001

Petre M, Minocha S, Roberts D (2006) Usability beyond the website: an empirically grounded e-commerce evaluation instrument for the total customer experience. Behav Inform Technol 25(2):189–203

Preece J, Rogers Y, Benyon D, Holland S, Carey T (1994) Human computer interaction. Addison-Wesley, Wokingham

Preece J, Rogers Y, Sharp H (2002) Interaction design: beyond human-computer interaction. Wiley, New York

Ramani K (2010) Sustainable design. J Mech Des 132:1–2

Rego H, Moreira T, Grarcia-Penalvo F (2010) Web-based learning information systems for web 3.0. In: al. MDLe (ed) WSKS 2010, Part 1, CCIS 111. Springer-Verlag, Berlin Heidelberg, pp 196–201

Rhodes JS (2000) Usability can save your company. http://webword.com/moving/savecompany.html. Accessed 1 May 2015

Sengers P, McCarthy J, Dourish P (2006) Reflective HCI: articulating an agenda for critical practice. In: CHI' 06, New York, pp 1683–1686

Sengers P, Beale R, Knouf N (2009a) Sustainable HCI meets third wave HCI: 4 themes. In: CHI 2009, Boston, pp 0–3

Sengers P, Boehner K, Knouf N (2009b) Sustainable HCI meets third wave HCI: 4 themes. In: CHI 2009b. Boston

Shackel B (2009) Usability – context, framework, definition, design and evaluation. Interact Comput 21:339–346

Sharma A, Lyer G, Mehrotra A, Krishnan R (2010) Sustainability and business-to-business marketing: a framework and implications. Ind Mark Manag 39:330–341

Sharp H, Rogers Y, Preece J (2011) Interaction design – beyond human-computer interaction. Wiley

Shaw G, Walters R, Kumar A, Sprigg (2015) A sustainability in infrastructure asset management. In: Proceedings of the 7th World Congress on Engineering Asset Management (WCEAM 2012), Springer, pp 525–534

Shneiderman B, Plaisant C (2010) Designing the user interface: strategies for effective human-computer interaction. Addison Wesley, Reading

Silberman MS, Tomlinson B (2010) Toward an ecological sensibility: tools for evaluating sustainable HCI. In: CHI 2010, Atlanta, pp 3469–3474

Smith P, Sharicz C (2011) The shift needed for sustainability. Learn Organ 18(1):73–86

Stapledon T, Shaw G, Kumar A, Hood D (2015) Understanding the business case for infrastructure sustainability. In: Proceedings of the 7th World Congress on Engineering Asset Management (WCEAM 2012), Springer, pp 535–543

Stelzer K (2006) Sustainability=Good Design Les Ateliers De Lethique 2:1–15

Te'eni D, Carey J, Zhang P (2007) Human computer interaction: developing effective organizational information systems. Wiley, New York

Thakurta R (2010) Management of requirement volatility – a study of organizational competency and how it is influenced by the project environment. J Inf Technol Manag XXI((n/a)):24–34

Thatchenkery T, Avital M, Cooperrider D (2010) Introduction to positive design and appreciative construction: from sustainable development to sustainable value. Adv Appreciative Inq 3:1–14

Usability First (2002) Introduction to user-centered design. http://www.usabilityfirst.com/about-usability/introduction-to-user-centered-design/. Accessed 1 July 2012

Vora P (1998) Human factors methodology for designing Web sites. In: Chris Forsythe EGJR (ed) Human factors and web development. Lawrence Erlbaum Associates, Mahwah, pp 153–172

Wals AEJ (2014) Sustainability in higher education in the context of the UN DESD: a review of learning and institutionalization processes. J Clean Prod 62:8–15

Wang L, Kwok JS, Tsang DC, Poon C-S (2015a) Mixture design and treatment methods for recycling contaminated sediment. J Hazard Mater 283:623–632

Wang L, Tsang DC, Poon C-S (2015b) Green remediation and recycling of contaminated sediment by waste-incorporated stabilization/solidification. Chemosphere 122:257–264

Weybrecht G (2010) The sustainable MBA – the manager's guide to green business. Wiley, Chichester

Wilson P, Borras J (1998) Lessons learnt from an HCI repository. Int J Ind Ergon 22(4–5):389–396. doi:10.1016/s0169-8141(97)00093-0

Young W, Tilley F (2006) Can businesses move beyond efficiency? The shift toward effectiveness and equity in the corporate sustainability debate. Bus Strateg Environ 15:402–415

Chapter 2
Usability and Human Computer Interaction (HCI)

Abstract Usability and HCI are becoming core aspects of the system development process to improve and enhance system facilities and to satisfy users' needs and necessities. HCI will assist designers, analysts and users to identify the system needs from text style, fonts, layout, graphics and color, while usability will confirm if the system is efficient, effective, safe, utility, easy to learn, easy to remember, easy to use and to evaluate, practical visible and provide job satisfaction to the users.

Adopting these aspects in the system development process, including the sustainable design will measure and accomplish users' goals and tasks by using a specific technology. Finally, designers should include these aspects in their agenda to enhance technology acceptance, performance and satiate users' necessaries.

2.1 Introduction

This discusses the value and the meaning of Human Computing Interaction (HCI) and its usefulness in designing a user interface or website. "Human Computer Interaction (HCI) is about designing a computer system that supports people so that they can carry out their activities productively and safely" (Preece et al. 1994, p. 1). HCI plays an important role in the development of computer systems and websites as it helps to develop "interactional techniques and to suggest where and in what situations these technologies and techniques might be put to best use" (Booth 1989, p. 6).

Thus, a commercial websites with effective HCI are likely to be more useful and profitable. HCI is a "very important concept in the system development process as it is about understanding and creating software and other technology that people will want to use, will be able to use, and will find effective when used. And the usability concept and the methods and tools to encourage it, achieve it, and measure it are now touchstones in the culture of computing" (Carroll 2002, p. xxvii). In addition, this chapter addresses the topic of Usability Evaluation, as usability "is concerned with both obtaining user requirements in the early stages of design, and with evaluating systems that have been built" (Booth 1989, p. 103).

There are various methodologies to create effective websites; these methodologies address detailed issues such as page design, typography, graphics, sound, navigation, and multimedia. However, they do not provide an adequate overall approach to HCI and usability.

2.2   User-Centered System Design

In order for computer-based systems to be widely accepted and used effectively, they need to be well designed via a "user-centered" approach. This is not to say that all systems have to be designed to accommodate everyone, but that computer-based systems should be designed for the needs and capabilities of the people for whom they are intended. In the end, users should not even have to think about the complexity of how to use a computer. For that reason, computers and related devices have to be designed with an understanding that people with specific tasks in mind will want to use them in a way that is seamless with respect to their work. Additionally, it is very important to "define style, norms, roles and even mores of human and computer relationship that each side can live with, as computers become more complex, smarter and more capable," and as we allow them to "take on autonomous or semi-autonomous control of more critical aspects of our lives and society"(Miller 2004, p. 34).

Systems designers need to know how to think in terms of future users' tasks and how to translate that knowledge into an executable system. This can be accomplished by establishing a good interface design to let the user interact and deal with the computer without any difficulties and to have more control of the system. Head (1999, p. 6) stated that good interface design "is a reliable and effective intermediary, sending us the right cues so that tasks get done – regardless of how trivial, incidental, or artful the design might seem to be".

Recently, as we know, user-centered design has become an important "concept in the design of interactive system[s]. It is primarily concerned with the design of sociotechnical systems that take into account not only their users, but also the use of technologies in users' everyday activities, it can be thought of as the design of spaces for human communications and interaction" (DePaula 2003, p. 219).

HCI "is recognized as an interdisciplinary subject" (Dix et al. 2004, p. 4). HCI needs input from a range of disciplines; for example, "computer science (application design and engineering of human interfaces), psychology (the application of theories of cognitive processes and the empirical analysis of user behavior), sociology and anthropology (interactions between technology, work, and organization), and industrial design (interactive products)". Therefore, HCI has "science, engineering, and design aspects" (Hewett et al. 1992).

2.3   Human Computer Interaction (HCI)

Before detailed consideration of the topic of Human Computer Interaction, two terms should be defined which are related to the development process: 'Interface' and 'Interaction'? According to Head, Interface is the "visible piece of a system that a user sees or hears or touches" (Head 1999, p. 4). Interaction is a more general term covering the users' activity. For instance, when the user types something by using the keyboard or clicks with a mouse, this activity is called interaction.

The general concepts of HCI apply to website design. Website designers have noticed that creating a "user friendly" site is important to maximize user response. However, designers "did[not] know any effective ways to discover what made a product user-friendly or how to design a product that was friendly" (McCracken and Wolfe 2004 p. 3). Designers often have a poor understating of HCI issues. Therefore, designers need to know how to think in terms of future users' needs, values, and supportable tasks and how to translate that knowledge into an executable system. This can be accomplished by establishing a good interface design to let the user interact and deal with the websites without any difficulties and to let the user have more control of the site.

Furthermore, in order to work effectively in the development process, HCI needs to be part of this process. According to Head, HCI has two critical dimensions in the development process: firstly, involving the user during the building and implementation of the new systems; secondly, evaluation studies about "cognitive and other behavioral factors that come into play when people interact with computers" (Head 1999, p. 9). These dimensions are consistent and mutually dependent, thus "the evaluation side of HCI becomes(s) a basis for decision making about design trade-offs during product development" (Head 1999, p. 9).

In the past, HCI experts tended to be consulted later in the design process, but most of the research found that this was a mistake. "The Interface is not something that can be plugged in at the last minute; its design should be developed integrally with the rest of the system. It should not just present a "pretty face"; but should support the tasks that people actually want to do, and forgive the careless mistakes" (Dix et al. 2004, p. 3). Thus, it is important to consider how HCI will fit into the overall design process for websites (see Fig. 2.1).

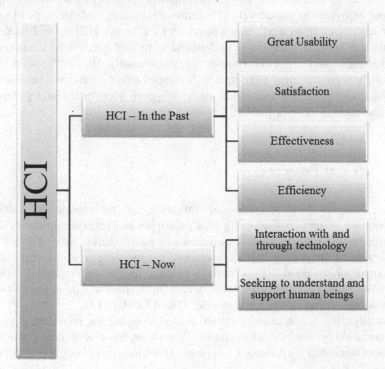

Fig. 2.1   HCI – past and now (Prepared by Tomayess Issa)

2.3.1   What Is HCI?

The term Human-Computer Interaction (HCI) was adopted in the mid-1980s as a means of describing this new field of study. "This term acknowledged that the focus of interest was broader than just the design of the interface and was concerned with all those aspects that relate to the interaction between users and computers" (Preece et al. 1994, p. 7).

HCI "is a discipline concerned with the design, evaluation, and implementation of interactive computing systems for human use and with the study of major phenomena surrounding them" (Preece et al. 1994, p. 7). Therefore, the reasons for studying HCI in the development process are to create interactive computer systems that are usable and practical as well (Head 1999).

The term HCI relates to several stages in the development process, including the design, implementation, and evaluation of interactive systems, in the "context of the user's task and work" (Dix et al. 2004, p. 4). The implementation of HCI can be perceived as an art as well as a science because it requires a comprehensive range of skills, including an understanding of the user, an appreciation of software engineering capabilities and application of appropriate graphical interfaces. "If we are to be recognized as developers with professional capabilities, as competent practitioners, then it is critical to understand what makes an application interactive, instructional and effective" (Sims 1997).

HCI "is concerned with the design of computer systems that are safe, efficient, easy, and enjoyable to use as well as functional" (Preece et al. 1993, p. 11). Vora (1998) describes a framework, which provides for effective HCI for websites, with the main task being to have a clear understanding of user needs: who the users are, and what their tasks and environments are. Additionally, HCI is "concerned not only with how present input and output technologies affect interaction, but also with the consequences of new techniques such as speech recognition and generation (input and output)" (Booth 1989, p. 5).

2.3.2   HCI as Process

HCI is a discipline focusing on design, evaluation, and implementation of interactive computer systems. By adopting HCI principles and practices in the development process, the system should be easy to use by people within their work settings. The purpose of integrating HCI techniques in the overall development process is that it incorporates good design "both in practice and in understanding", and to achieve this goal, HCI addresses "what occurs on the human side of interaction as well as what happens on the machine side" (Head 1999, p. 12).

Basically, HCI is concerned with two issues: studying the relationship and the communication between the human and the computer, and discovering the methods for "mapping computing functions to human capabilities and effectively using input and output techniques so that computers and users have more seamless interactions" (Head 1999, p. 12). HCI places a special emphasis on "creating and applying user-centered design techniques as well as using iterative usability testing methods" (Head 1999, p. 13).

Consequently, the machine [Computer] side involves several relevant issues including "computer graphics, operating systems, programming languages and development environments." While on the human side, "communication theory, graphic and industrial design disciplines, linguistics, social science, cognitive psychology, and human performance are relevant" (Hewett et al. 1992).

2.3.3 Relationship Between the HCI and Human Dialogue

HCI is the study and theory of the interaction between humans and complex technology and is concerned with how current input and output technologies affect interaction, and the situations in which these technologies and techniques might be put to best use. Therefore, the relationship between HCI and human dialogue may be summarized as follows: (Booth 1989, p. 54–55).

- Human Computer interaction, like human dialogue, is a form of communication where a degree of understanding can be achieved. Admittedly, this understanding may be limited in some respects, but if designed properly, a computer system will do as its user wishes, provided the user knows what is possible and how to give commands.
- Communications requires agreement on the terms used in the dialogue. When humans successfully communicate, they usually have a shared understanding of the words' used and the concepts to which they refer. This is also true of human computer communication. When a user gives commands to a system, then the system must have an understanding of these commands if the interaction is going to succeed.
- Communications requires agreement, not only upon the terms and concepts used, but also upon the context of the communication.

For example, if two people are speaking to one another, then there needs to be an agreed understanding of what they are speaking about. To illustrate this point further, let us consider an example where two individuals do not agree on the context of their conversation. Two people are sharing a car to travel to a conference. They stop at a garage for fuel and to check the car tyres. Bill is putting air into the tyres when Fred asks, "How's the pressure?" Bill replies, "Not too good, the boss keeps getting on to me." Fred explains, "Sorry I meant the car tyre pressure, but how's work anyway?" (Booth 1989, p. 55). In this example, we understand that Fred and Bill do not share a common context for their brief exchange. "In their separate contexts, the necessary link of work and the context of car maintenance, some of the words can have different meanings (i.e." Pressure") and the result is a failure in the dialogue between the two individuals" (Booth 1989, p. 55).

This sort of dialogue failure can also occur in human-computer communications. For example, "consider a user of a word processing system who issues a command to print the document that is currently being edited." Following the printing process, "the user issues a command for the system to re-display the document on the screen, but instead nothing happen." The system, "upon receiving the first command changed to the printing mode, but did not adequately inform the user who was unaware of the change in context and the subsequent legality of some of the commands." The lesson to be learned is "that those involved in communication assign [meaning] to symbols and terms depend[ing] upon the context in which they are communicated" (Booth 1989, p. 55).

The previous two examples reveal that perspective is not only important in conversation between humans, but is also a considerable factor in human-computer dialogue. To sum up, HCI is similar to human dialogue, as it is a form of communication where a degree of understanding is achieved. There must also be agreement between individuals involved in the process of communication on the meaning of the symbols and terms used. The context of the dialogue is also important, as it is the context that dictates the meanings of some of the symbols and terms used.

2.3.4   Goals of HCI

The goals of HCI are to produce usable and safe systems, as well as functional systems. These goals can be summarized as safety, utility, effectiveness, efficiency, and appeal. These goals focus on the services that the system provides, how quickly the tasks can be achieved, and ensuring that users like the system. In general, usability is an essential concept in HCI and is concerned with making systems easy to learn, easy to use, and with limiting error frequency and severity. To establish a simple system with good usability, the HCI specialists need to be aware of the following issues (Preece et al. 1994, p. 15):

- Understand the factors such as organizational, social, and psychological factors that determine how people operate and make use of computer technology effectively.
- Develop tools and techniques to help designers ensure that computer systems are suitable for the activities for which people will use them.
- Achieve efficient, effective, and safe interaction in terms of both individual Human Computer Interaction and group interaction.

These needs should be considered very carefully at the design stage, as most of the users should not have to change radically to 'fit in' with the system; rather, the system should be designed to match their requirements.

2.3.5   Purpose of HCI

The purpose of HCI is to design a computer system to match the needs and requirements of the users. The HCI specialists need to think about the above factors in order to produce an outstanding system. To achieve the goals of HCI, a number of approaches can be utilized. These approaches need to be studied very carefully in order to develop a system, which provides the user with productivity and efficiency. These approaches are: (Preece et al. 1994, p. 46–47)

- Involving the user: (involve the user as much as possible so that s/he can influence the system design).
- Integrating different kinds of knowledge and expertise: (integrate knowledge and expertise from the different disciplines that contribute to HCI design).
- Making the design process iterative: (testing can be done to check that the design does indeed meet users' requirements).

From the above, it was learned that HCI design should be user-centered, integrate knowledge from different disciplines, and be highly iterative. In addition, it is important to undertake effective usability evaluation. This will provide feedback regarding negative and positive aspects of prototypes.

It is important that the way in which people interact with computers is intuitive and clear. However, designing appropriate HCI is not always straightforward, as the many poorly designed computer systems testify. One of the challenges of HCI design is to keep abreast of technological developments and to ensure that these are harnessed for maximum human benefit.

The goal of this research is to develop a framework for rapid, integrated, incremental systems development that enables a group of designers and users working together to produce a friendly, effective and efficient website. Two terms – Interaction and Interactivity – need to be defined in order to understand how the user can communicate with the system to accomplish his/her goals.

2.3.6   Interaction and Interactivity

According to Dix, "Interaction involves at least two participants: the user and the system. Both are complex, as we have seen and are very different from each other in the way that they communicate and view the domain and the task. The interface must, therefore, effectively translate between them to allow the interaction to be successful" (Dix et al. 1998, p. 104).

Users can interact with computer systems in a variety of ways. At the lowest level is batch input, in which the user provides all the information to the computer at once and leaves the machine to perform the task. This approach is called indirect interaction. An approach which involves a real-time interaction between the users and the computer is called direct interaction, as a dialogue between the user and computer will be established and at the same time will provide feedback and control right through to achieving the task.

The study of interaction can help both the HCI specialists and the users simultaneously; for example, analysis of interaction will help HCI specialists to understand exactly what is going on in the interaction, and identify the likely root of difficulties. It can compare different interaction styles and take into account the interaction problems. On the other hand, the users are able to achieve their goals successfully. These goals relate to the particular application domain i.e. an "area of expertise and knowledge in some real-world activity" (Dix et al. 1998, p. 104). The user interacts with the system for a specific reason – i.e. to perform a task, in turn to achieve the goal, which was (for instance) the reason behind visiting a particular website. So the goal is "the desired output from a performed task" while the task is an "operation to manipulate the concepts of a domain" (Dix et al. 1998, p. 104).

To understand the interaction concept, Norman's model of interaction can be utilized (see Fig. 2.2) (Norman 1986). This model may be considered as a cycle between execution and evaluation, and these two stages can be subdivided into seven steps. The user begins the interactive cycle by defining the goal and the tasks in order to achieve his/her objectives. The user will define his/her goal by using the input mechanisms, so the task must be "articulated within the input language" (Dix et al. 1998, p. 107). Then the input language will be translated into the system language (known by Norman as Core Language). Later, the system then "transforms

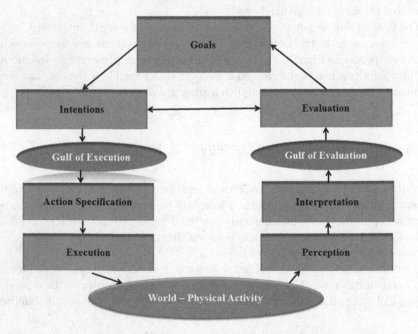

Fig. 2.2 Norman's interaction model (Adopted from Norman (1986). Prepared by Tomayess Issa)

itself as described by the operation translated from the Input; therefore, the execution phase is complete" (Dix et al. 1998, p. 107). If the system responds to the user task in an appropriate manner to achieve the goal, then the interaction has been successful between the user and the system; otherwise, the user must "formulate a new goal and repeat the cycle" (Dix et al. 1998, p. 106).

Next, the evaluation phase begins, as the system will be in the new state and must communicate to the user the current values of the system since "attributes are rendered as concepts or features of the output" (Dix et al. 1998, p. 107). Thus, the user can see the consequences of the task s/he initiated.

Finally, is up to the user to interpret the output and to match the results of the "interaction relative to the original goal" (Dix et al. 1998, p. 107). At this stage, the evaluation phase has ended as has the interactive cycle. A new cycle may then commence.

Norman's model is very useful as a means to understand the principles behind the interaction framework. This model allows the user to define his/her goals firstly and then will let them interact with the system to accomplish these goals. However, other researchers suggest that Norman's model considers only the "system as far as the interface, and is only focusing on the user's view of the interaction" (Dix et al. 1998, p. 106). A more complex approach is needed.

The second way in which to discuss the users' communication with the system is interactivity. Interactivity can be defined in general terms as "the facility for individuals and organizations to communicate directly with one another regardless of distance or time" (Ghose and Dou 1998, p. 30). For instance, in an educational context, interactivity "refers to the activity between two organisms – which are learner and the computer" (Jonassen 1998, p. 97). In the context of HCI, "Interactivity is the defining feature of an interactive system. This can be seen in many areas of HCI such as recognition rate for speech, recognition, and 'feel' of a WIMP environment element: windows, icons, menus, pointers, dialog boxes, and buttons" (Dix et al. 1998, p. 136). This process is iterative with a sequence of steps and procedures followed by the user to interact with the machine (or system) to further his/her goal.

2.3.7   Factors in HCI Design

To achieve a safe and user-friendly system, the HCI specialists need to consider the main issues and factors involved in interaction and interactivity, and hence in HCI design (see Fig. 2.3). These factors can be divided into (Preece et al. 1994, p. 31):

- *Organizational factors* (training, job design, politics, roles, work organization);
- *Environmental factors* (noise, heating, lighting, ventilation);
- *Health and Safety factors* (stress, headaches, musculo-skeletal disorders);
- *The User* (motivation, enjoyment, satisfaction, personality, experience level);
- *Comfort Factors* (input devices, output displays, dialogue structures, use of color, icons, commands, graphics, natural language, 3-D, user support materials, multi-media);

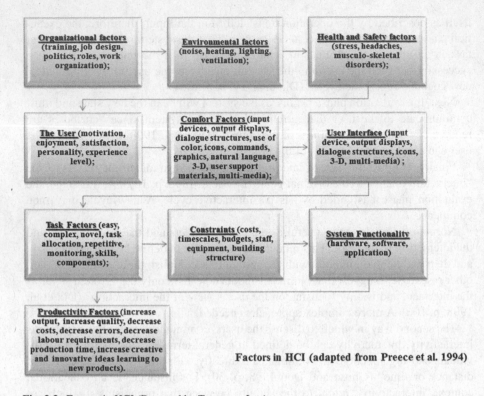

Factors in HCI (adapted from Preece et al. 1994)

Fig. 2.3 Factors in HCI (Prepared by Tomayess Issa)

- *User Interface* (input device, output displays, dialogue structures, icons, 3-D, multi-media);
- *Task Factors* (easy, complex, novel, task allocation, repetitive, monitoring, skills, components);
- *Constraints* (costs, timescales, budgets, staff, equipment, building structure);
- *System Functionality* (hardware, software, application);
- *Productivity factors* (increase output, increase quality, decrease cost, decrease errors, decrease labor requirements, and decrease production time, increase creative and innovative ideas leading to new products).

Many factors are involved, therefore, during the development process; disagreement can arise between ways to address each of these factors depending on various aspects of the system development context, such as product, team members, users, and company. According to Head (1999, p. 33) "making careful trade-offs between these numerous factors, while supporting design principles and approaches, remains a challenge of the HCI field". Consequently, most designers support involvement of the user in the design process from the beginning to reduce conflicts during the development stage.

- Allocate, users, analysts, and designers (internal and external) to identify that the website design is practical.
- There are many specific issues that need to be taken into consideration when designing website pages, such as
  - Text style
  - Fonts
  - Layout
  - Graphics
  - Color.

Fig. 2.4 HCI step in the New Participative Methodology for Marketing Websites (NPMMW) – Issa 2008 (Prepared by Tomayess Issa)

Finally, Issa (2008) indicates that HCI is essential in the system development system. HCI will allocate users, analysts, and designers (internal and external) to identify that the website design is practical. Many specific issues need to be taken into consideration when designing website pages, such as text style, fonts; layout, graphics, and color (see Fig. 2.4).

2.4 What Is USABILITY?

Usability refers to the "quality of the interaction in terms of parameters such as time taken to perform tasks, number of errors made, and the time to become a competent user" (Benyon et al. 2005, p. 52). Alternatively, Usability "is a **quality attribute** that assesses how easy user interfaces are to use. The word "usability" also refers to methods for improving ease-of-use during the design process" (Nielsen 2003). The usability evaluation stage is an effective method by which a software development team can establish the positive and negative aspects of its prototype releases, and make the required changes before the system is delivered to the target users. Usability evaluation is about observing users to "see what can be improved, what new products can be developed" (McGovern 2003). It is "based on human psychology and user research" (Rhodes 2000). HCI specialists "observe and talk with participants as they try to accomplish true-to-life tasks on a site (or system), and this allows them to form a detailed picture of the site as experienced by the user" (Carroll 2004).

From the user's perspective, usability is considered a very important aspect in the development process as it can mean the difference between performing and completing a task in a successful way without any frustration. Alternatively, if usability

Fig. 2.5 Usability
(Prepared by Tomayess
Issa)

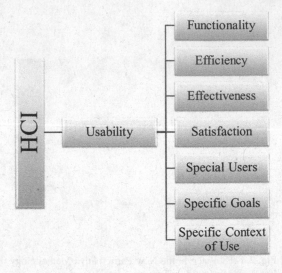

is not highlighted in website design, then users will become very frustrated working with it (see Fig. 2.5). For example, according to Nielsen (2003), people will leave the website: (a) if is difficult to use; (b) if the users get lost on a website; (c) the information is hard to read; (d) it does not answer users' key questions; (e) and lastly, if the homepage fails to define the purpose and the goals of the website. "Usability rules the web. Simply stated, if the customer cannot find a product, then s/he will not buy it. In addition, the web is the ultimate customer-empowering environment. S/he who clicks the mouse gets to decide everything. It is so easy to go elsewhere; all the competitors in the world are but a mouse-click away" (Nielsen and Mack 1994, p. 9).

Usability is a critical issue for websites as it improves competitive position, improves customer loyalty, and drives down costs (Rhodes 2000). Therefore, if usability is highlighted in website design, it will keep the organization in a powerful position compared with their competitors, as "Usability = simplicity = user satisfaction = increased profits" (Rhodes 2000).

2.4.1   Concepts of Usability

To understand fully the concepts behind the term "usability," we need to realize that usability is not "determined by just one or two constituents, but is influenced by a number of factors" which interact with "one another in sometimes complex ways" (Booth 1989, p. 106). Eason (1984) has suggested a sequence of models (see Fig. 2.6) that clarify what these variables might be. Figure 2.6 displays the relationship between independent (task, user, and system characteristics) and dependent variables (User Reaction) with each variable having specific requirements and needs.

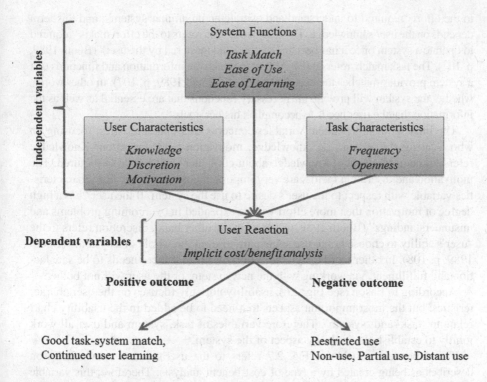

Fig. 2.6 Eason's causal framework of usability (Adopted from Eason (1984). Prepared by Tomayess Issa)

First, task characteristics are divided into frequency and openness. The frequency term refers to "the number of times any particular task is performed by a user" (Booth 1989, p. 107). If users perform a task infrequently, then help and assistance should be available via the interface so that users know which step must be taken next to accomplish the task. On the other hand, if users perform a task frequently, then it will be easier for him/her to remember the steps, which are required in order to accomplish the task.

The openness term refers to the "extent to which a task is modifiable" (Booth 1989, p. 107). This means that the information needs of the user are variable and the task must "be structured to allow the user to acquire a wide range of information." According to Eason (cited in Booth (1989)), the user information needs should be fixed. If this is the situation at that time "the task need not be open and flexible, as the same information is required each time the task is performed" (Booth 1989, p. 107).

The system function is described as being the most important concept under the causal framework for usability. The main concept of this variable is to improve the usability under the development process. To achieve this, the system function must address the three major system variables carefully within the development process. These are ease of learning, ease of use and task match. The ease of learning term refers to the effort "required to understand and operate an unfamiliar system"; and this term depends on the user's knowledge. The ease of use term refers to the effort that is "required to operate a system once it has been understood and mastered by the user" (Booth 1989, p. 107). The task match, refers to the "extent to which the information and functions that a system provides matches the needs of the user" (Booth 1989, p. 107); in other words, whether the system will provide the necessary functions that are essential as well as the information that the user needs to accomplish his/her goals.

The final set of independent variables concerns user characteristics, focusing on who is using the system, i.e. knowledge, motivation, and discretion. Knowledge refers to the user's level of knowledge about computers and the tasks required. The motivation and discretion factors are very important concepts in the user characteristics variable with respect to the user's desire to use the system. If the user "has a high degree of motivation then more effort will be expended in overcoming problems and misunderstandings" (Booth 1989, p. 108). On the other hand, discretion refers to the "user's ability to choose not to use some part, or even the whole of a system" (Booth 1989, p. 108). In other words, high discretion means that there needs to be satisfaction and fulfillment, via working with the new system, or the user will not bother.

According to Eason (see Fig. 2.6), usability not only focuses on the user characteristics, but the most important aspects that need to be added in the usability chart relate to 'task' and 'system'. Therefore variables of task, system and user all work jointly to establish the usability aspect of the system.

The dependent variable in Fig. 2.7 refers to the user's reaction, which Eason describes as being created by a type of cost-benefit analysis. Therefore, this variable focuses on the negative and positive outcomes of adopting the new system. Positive outcomes will lead to success of the system, while the negative outcomes will lead to suspension and discontinuation of the system. In other words, the user "accumulates a knowledge base of task-system connections as the system is used in a sequence of task episodes. The emerging strategy for use may represent a positive outcome in which the

Fig. 2.7 A re-iteration of Eason's (1984) interacting task, system and user variables (Adopted from Booth 1989, p. 109) (Prepared by Tomayess Issa)

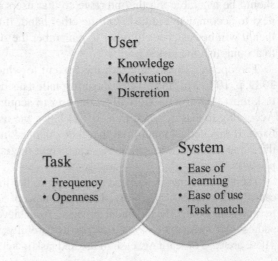

user locates and uses appropriate system functions for every new task and progressively masters the system. The reverse scenario occurs when negative outcomes prevail and use of the system is discontinued. Eason points out, based on his field studies, that under realistic conditions the user appears to approach a state of equilibrium where further learning about the system is minimized" (Lowgren 1995, p. 5).

2.4.2   Usability Criteria

Various principles need to be followed in order to support usability, making systems easy to learn and easy to use. These principles are (Dix et al. 1998, p. 162 and Nielsen 2003):

- *Learnability:* by which new users can begin effective interaction and achieve maximal performance;
- *Flexibility:* the multiplicity of ways the user and system exchange information;
- *Robustness:* the level of support provided to the user in determining successful achievement and assessment of goals;
- *Efficiency:* once the user learns about the system,[the speed with which s/he] can perform the tasks;
- *Memorability:* how easily the user will remember the system functions, after a period time of not using it;
- *Errors:* "How many errors do users make, how severe are these errors, and how easily can they recover from the errors?" (Nielsen 2003);
- *Satisfaction:* how enjoyable and pleasant is it to work with the system?

These principles can be applied to the design of an interactive system in order to promote its usability. Therefore, the purposes behind adopting these principles are to give more assistance and knowledge to system developers (and the users) regarding the system design. Alongside the above principles, an important key additional factor is Utility. Utility refers to the functionality so users can "do what they need or want to do" (Preece et al. 2002, p. 16). In other words, "does it do what users need?" (Nielsen 2003). For that reason, usability and utility are equally important in the development process and they need to be integrated.

2.4.3   Usability Specifications

Once the designer has gathered and analyzed information about the tasks, problems and steps to work with the proposed system, the next step is to answer the question: How will we know if the interface is usable? This is laid out in a usability specification.

A usability specification defines the measure of success of a computer system or website and serves as an indicator about whether or not the development of the website is on the right track. A usability specification should be developed during the first stage of the development process and monitored "at each iteration", to determine whether the "interface, is, indeed, converging toward an improved, more usable design" (Hix and Hartson 1993, p. 222). Usability specifications should lay out explicitly how usability will be evaluated and can be divided into two sections:

- *Performance Measures:* are directly observable by watching a user complete a task within a specific time. This includes monitoring the number of errors and time needed to accomplish the task. These types are 'quantifiable measures' which means that they can be communicated with numbers. For example "you can count the number of minutes it tasks a user to complete a task or the number of negative comments that occur" (McCracken and Wolfe 2004, p. 53).
- *Preference Measures:* give an indication of a "user's opinion about the interface which is not directly observable" (McCracken and Wolfe 2004, p. 53). Preference measures can be determined by using questionnaires or interviews.

Usability specifications are needed to determine when the iteration of prototypes has produced a system with sufficient usability. Therefore, without usability specifications, the key factors that "generally determine an end to the iterative refinement process are when developers run out of time, patience, and/or money" (Hix and Hartson 1993, p. 243). Usability specifications are very important to the development process since they define "a quantifiable end to the seemingly endless iterative refinement process" (Hix and Hartson 1993, p. 242).

Lastly, Issa (2008) confirm that usability is a core step in the system development process as usability will allow users, analysts, and designers (internal and external) to confirm that the website design is efficient, effective, safe, utility, easy to learn, easy to remember, easy to use and to evaluate, practical, visible and provide job satisfaction (see Fig. 2.8).

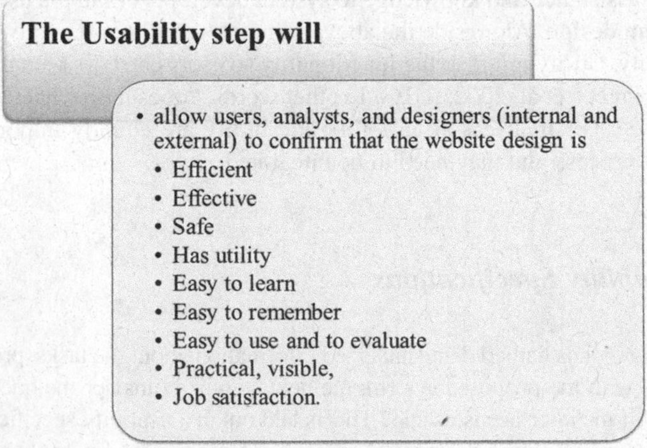

The Usability step will

- allow users, analysts, and designers (internal and external) to confirm that the website design is
  - Efficient
  - Effective
  - Safe
  - Has utility
  - Easy to learn
  - Easy to remember
  - Easy to use and to evaluate
  - Practical, visible,
  - Job satisfaction.

Fig. 2.8 HCI step in the New Participative Methodology for Marketing Websites (NPMMW) – Issa 2008 (Prepared by Tomayess Issa)

2.5 Conclusion

This chapter has outlined the basic concepts involved in Human Computer Interaction and usability in the system development process. These considerations are very useful to the business community in line to increase the efficiency of their staff, and thus, their profits. Currently, HCI and usability are needed in any design, including sustainable design to recognize the new smart technology and portable device needs from designers and users perspective. Therefore, designers should integrate and combine HCI and usability in their agenda design, including sustainable design, to enhance new smart technology and portable devices performance and facilities, and to satisfy users' needs.

References

Benyon D, Turner P, Turner S (2005) Designing interactive systems: a comprehensive guide to HCI and interaction design, 2nd edn. Pearson Education Limited, Edinburgh

Booth P (1989) An introduction to human-computer interaction. Lawrence Erlbaum Associates Publishers, Hove/East Sussex

Carroll JM (2002) Human-computer interaction in the new millennium. Addison-Wesley, New York

Carroll M (2004) Usability testing leads to better ROI. http://www.theusabilitycompany.com/news/media_coverage/pdfs/2003/NewMediaAge_270303.pdf. Accessed 1 Sept 2014

DePaula RA (2003) New era in human computer interaction: the challenges of technology as a social proxy. In: Latin American conference on HCI, ACM international conference proceeding series, Rio de Janeiro, Brazil, pp 219–222

Dix A, Finlay J, Abowd G, Beale R (1998) Human-computer interaction, 2nd edn. Pearson Education Limited, Harlow

Dix A, Finlay J, Abowd G, Beale R (2004) Human-computer interaction, 3rd edn. Pearson Education Limited, Harlow

Eason KD (1984) Towards the experimental study of usability. Behav Inform Technol 3(2):133–143

Ghose S, Dou W (1998) Interactive functions and their impacts on the appeal of internet presence sites. J Advert Res 38(2):29

Head AJ (1999) Design wise. Thomas H Hogan Sr, Medford

Hewett T, Baecker R, Card C, Carey T, Gasen J, Mantei M, Perlman G, Strong G, Verplank W (1992) Human-computer interaction. ACM SIGCHI curricula for human-computer interaction. http://old.sigchi.org/cdg/. Accessed 1 Mar 2011

Hix D, Hartson HR (1993) Developing user interfaces: ensuring usability through product & process. Wiley, New York

Issa T (2008) Development and evaluation of a methodology for developing websites. PhD thesis, Curtin University, Western Australia. http://espace.library.curtin.edu.au/R/MTS5B8S4X3B7S BAD5RHCGECEH2FLI5DB94FCFCEALV7UT55BFM-00465?func=results-jump -full&set_entry=000060&set_number=002569&base=GEN01-ERA02

Jonassen DH (1998) Instructional designs for microcomputer courseware. L. Erlbaum Associates, Hillsdale

Lowgren J (1995) Perspectives on usability. Department of Computers and Information Science, Linkoping University, Sweden. http://www.ida.liu.se/labs/aslab/groups/um/publications/R-95-23. PDF. Accessed 19 Apr 2004

McCracken DD, Wolfe RJ (2004) User-centered website development a human-computer interaction approach. Pearson Education Inc., New Jersey

McGovern G (2003) Usability is good management. http://www.gerrymcgovern.com/nt/2003/nt_2003_04_07_usability.htm. Accessed 1 May 2015

Miller CA (2004) Human-computer etiquette: managing expectations with intentional agents. Commun ACM 47(4):31–34

Nielsen J (2003) Usability 101. http://www.useit.com/alertbox/20030825.html. Accessed 1 May 2015

Nielsen J, Mack RL (1994) Usability inspection methods. Wiley, New York

Norman DA (1986) Seven-stage model of (individual) interaction. Department of Computer and Systems Sciences, Lulea University. Bai Guohua, Sweden

Preece J, Rogers Y, Keller L, Davies G, Benyon D (1993) In: Preece J (ed) A guide to usability "human factors in computing". Addison-Wesley, Wokingham

Preece J, Rogers Y, Benyon D, Holland S, Carey T (1994) Human computer interaction. Addison-Wesley, Wokingham

Preece J, Rogers Y, Sharp H (2002) Interaction design: beyond human-computer interaction. Wiley, New York

Rhodes JS (2000) Usability can save your company. http://webword.com/moving/savecompany.html. Accessed 1 May 2015

Sims R (1997) Interactivity: a forgotten art?. http://www.gsu.edu/~wwwitr/docs/interact/. Accessed 1 May 2015

Vora P (1998) Human factors methodology for designing web sites. In: Chris Forsythe EGJR (ed) Human factors and web development. Lawrence Erlbaum Associates, Mahwah, pp 153–172

Chapter 3
User Participation in the System Development Process

Abstract User participation in the system development process is crucial and vital to ensure if user interfaces, devices including website are successful and easy to learn and implement as user participation will improve and enhance performance and increase user acceptance and satisfaction. User participation will encourage users to participate in decision-making and actions during the system development process. The user participation rational will reduce the time taken by designers in various stages from implementation, testing, evaluation, and training, since users will become more aware behind the new design. This chapter aims to discuss the importance of user participation in the system development and sharing with the readers the why, how and when we need to involve participants in the design process.

3.1 Introduction

This chapter focuses on users, their work, and their environment and the reasons for involving them in the design process. Participation role in the system development process is crucial and critical to ensure if the design process will be successful or unsuccessful. In general, if designers manage to work very closely with the users to produce new smart technology or portable devices, then less time will be required in the implementation, testing and training stages, and consequently, the user will work with the new devices, with less frustration and dissatisfaction. This chapter is organized as follows: What is Participation, How We Know Our Users and Conclusion.

3.2 What Is Participation?

Participation is "A process in which two or more parties influence each other in making plans, policies or decisions, it is restricted to decisions that have future effects on all those making the decisions or on those represented by them" (Mumford 1995, p. 12). It can also be defined (in the context of systems development practices) as the "extent to which the user engages in systems analysis activities such as project definition and logical design decisions" (Doll and Torkzadeh 1989, p. 1154). Furthermore, user participation is defined as the "behaviors, assignments, and

T. Issa, P. Isaias, *Sustainable Design*, DOI 10.1007/978-1-4471-6753-2_3

activities that users or their representatives perform during the information system development" (Hartwick and Barki 1994, p. 441). A high level of user participation is likely to enhance user "ownership" of, or identification with the resulting system – in this sense "'user involvement' refers to the set of all such user subjective attitudes toward, or psychological identifications with, information systems and their development" (Kappelman 1995, p. 70). However, the term 'user involvement' can also refer to a low level of participation, where users have little power to influence decisions.

This research focuses on "user participation" not "user involvement" as the former term implies a role for the users which is more powerful and influential in the development process, especially in website design, as the user will be actively engaged throughout the development process. This will assist the user to accept and comprehend the system. Participation is more "effective when an individual's desire or "motivation to participate" is in congruence with perceptions of actual involvement" (Doll and Torkzadeh 1991, p. 443). Decisions about the role of the user need to take into consideration that users are "becoming more knowledgeable and active in defining their information requirements" (Doll and Torkzadeh 1989, p. 1154).

This research distinguishes between two types of users: end-users (internal to the client organization) and client-customer users (external). End-users (Internal) are the real users in the client organization, who test and evaluate the website and use it to respond to the client-customer's queries. The client-customer users (external) are those who interact with the website to accomplish their goals such as purchasing goods or services from the client organization. It is important to understand the needs, desires, and characteristics of both types of users. To date, most designers of websites have "assumed that their users had the same background and expectations that they did"; therefore, "the more you know about your users and their work, the more likely it is that you will develop a usable and successful website" (McCracken and Wolfe 2004, p. 37). These two types of users (see Fig. 3.1) should both participate in the development process under the methodology developed during this research, to make sure that the website meets the requirements of end-users, client-customers, and designers simultaneously. The purpose behind this participation has various benefits: (1) to reduce the time in the implementation and testing stages; (2) to familiarize the end-users and client customers with the new system before the implementation; (3) and provide job satisfaction and meet the task effectiveness needs of the end-users and client-customers.

User participation assists system development by providing a "more accurate and complete assessment of user information requirements, providing expertise about the organization the system is to support, expertise usually unavailable within the information systems group, avoiding development of unacceptable or unimportant features and importing user understating of the system" (McKeen et al. 1994, p. 427–428). Tait and Vessey stated that participation "reduces the risk of system failure in complex projects" (cited in (Amoako-Gyampah and White 1993, p. 2)). Therefore, in order to make the system more successful, participation needs to be an integral part of "the design and implementation process" (Tait and Vessey 1988, p. 91), not just a convenient add-on.

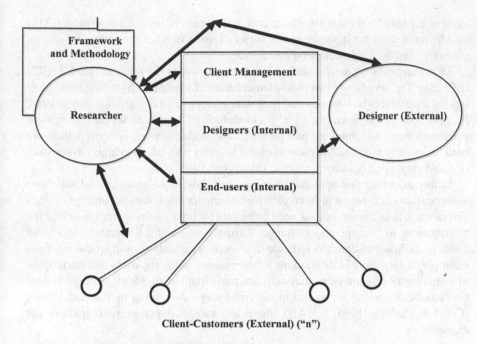

Fig. 3.1 Users (end-user and client-customers) (Prepared by Tomayess Issa)

Participation in the development process can be "viewed as "sharing" in decision making or engaging in activities" (Doll and Torkzadeh 1989, p. 1155), and to determine "information requirements by encouraging users and other to indicate what they do and what information they need to do it" (Hepworth et al. 1992, p. 122). Research has shown that user participation in system design will greatly assist in producing a successful system. It results in less time in the implementation and testing stages as users are more knowledgeable about the system.

The user's participation is very important since the lack of "user involvement as the chief reason IS projects fail" (Engler 1996, p. 3), and "developing an information system without user participation tends to result in the delivery of systems that fail to meet the users' needs" (Hawk and Dos Santos 1991, p. 317). After reviewing the role of user participation in different types of projects, Hirschheim asserts "more user participation was undertaken by organizations when the systems were complex" (cited in (Amoako-Gyampah and White 1993, p. 2).

User participation should be introduced in the development process to ensure that the system is successful and easy to implement as user participation may lead "to improved system quality as well as increased user acceptance, reflected in increased use of and satisfaction with the system" (Baroudi et al. 1986, p. 233). In addition, it will decrease resistance and increase acceptance of planned change (Baroudi et al. 1986). User participation will change "the attitude of user towards data processing and vice versa" (Doll 1987, p. 27).

Research and experience have shown that to run a successful application development process without any frustrations and dissatisfaction, the designer needs to involve the users, set clear objectives and recognition of organization factors. This will help the designer incorporate the views of users in all of the following stages: planning, design, implementation and testing.

To implement such an approach, a designer may adopt, for instance, the ETHICS (Effective Technical and Human Implementation of Computer Systems) methodology, as it considers both human and technical factors when designing a new system. In other words, this is known as a "socio-technical" approach, which "recognizes the interaction of technology and people, and produces work systems which are both technically efficient, have social characteristics which lead to high job satisfaction and create high quality products" (Mumford 1995, p. 2).

Before adopting this approach, a designer needs to understand, and take into account, that each user will have different characteristics, such as interest, values and needs. These considerations need to be met by both parties – employee and the management to "accept major change willingly and enthusiastically" (Mumford 1995, p. 2). Some researchers indicate that some organizations will let the management play a large role in developing a new system, while the users will participate in a small way, or sometimes they will not participate at all. Hence, user participation can be at various levels and in different ways. According to Tait and Vessey (Cited in (Saleem 1996, p. 147)), there are various types of participation, for example:

- *No participation:* users are not invited to participate;
- *Symbolic participation:* user input is sought but ignored;
- *Participation by advice:* users are consulted;
- *Participation by weak control:* users may have sign-off responsibility;
- *Participation by doing:* users are members of design team:
- *Participation by strong control:* users may pay for the system development.

The use of options involving little user participation will create numerous problems for the users as well as the management, as users will most likely find that this system is not meeting their needs, desires, and is very hard to cope with. This may lead to "serious morale problems" (Mumford 1995, p. 2) resulting in reduced job satisfaction, low efficiency "low commitment to the system, together with increased resistance to any future change" (Mumford 1995, p. 2).

3.2.1   Change Processes

To be successful and meet user requirements, the development of a new system requires a number of "change process" aspects to be considered by the designer, user and management simultaneously. These aspects are objective setting and attainment; adaptation; integration; and stabilization.

- *Objective Setting and Attainment:* this should involve all the groups (not only the senior management) from an organization who intend to use the system. Each group (or every individual) will have special interests and values. Consequently, designing a system for today and the future needs to involve various sessions of brainstorming between the users to exchange opinions and views to enhance the system. Today "non-technical users are familiar with, and knowledgeable about, the advantage and disadvantages of technical systems" (Mumford 1995, p. 6). Users are "becoming more sophisticated and as they do so, their expectations and behaviors are changing. Don't get caught designing for yesterday's audience – stay on the cutting edge with this kind of research so that you can design for tomorrow's audience!" (Sheridan 1999). Moreover, these groups are "able to make informed choices on the hardware and software that will best meet their needs" (Mumford 1995, p. 6).

- *Adaptation:* this process is "moving from one kind of technical and organizational structure and state to another, and the means by which this change is assisted to take place smoothly and successfully" (Mumford 1995, p. 7). Adaptation occurs in the implementation phase of the new system. The adaptation needs to address issues such as values, interests, attitudes, motivations and the conflicts between the groups who are working together to implement a new system. Therefore, support and assistance needs to be provided from the top management to understand and study any potential conflicts between groups of users. This step is very significant to reduce any struggle between the groups and to certify that the system is running smoothly, according to the users' needs.

- *Integration:* "is the action taken, once the system has been designed and is being implemented, to ensure a new situation reaches a state of equilibrium" (Mumford 1995, p. 7). The purpose behind integration is to gather different aspects such as task, technology, people and organizational environment into a valuable relationship between themselves. The relationship between these aspects should be stable and capable of adoption. Organizations should respond directly to all the changes which occur in their environment "while at the same time either maintaining a state of equilibrium or being able to make adjustments which restore equilibrium if internal relationships are distributed" (Mumford 1995, p. 8). Introducing a new technology to the above aspects (task, technology, people and organizational environment) will bring a new relationship between them, which should integrate "both opportunities and constraints" (Mumford 1995, p. 8). Since tasks are influenced by technology, the task structure of "functions or departments using the system will be altered" (Mumford 1995, p. 8). New tasks will have new demands; therefore, in this scenario, job satisfaction will be affected, as new tasks will have new demands and requirements that will produce negative or positive feedback. Consequently, technology, people and tasks will interact with the environment to provide a new structure "for the achievement of the organization's objectives and interaction may start the looping process again by making new demands of technology" (Mumford 1995, p. 8). Thus, integration requires adaptation in order to produce a good relationship between technology, people, tasks and organizational structure.

- *Stabilization:* this is the last step in the change process. Stabilization requires that "once new patterns of behaviour have been successfully initiated; they must be established and reinforced" (Mumford 1995, p. 6). This means that the relationship between the aspects (task, technology, people and organization) should incorporate the new patterns of task performance, which is required by the system to ensure that they meet the values and interests of groups who are involved.

In summary, designers need to take into account the above change processes during the development process of a new system, and these changes should be considered from the human perspective, not from the technical aspect. This means that user participation should be a priority from the beginning, involving the user in all stages of the process from planning to implementation. This action will achieve two desirable outcomes: a successful system and job satisfaction.

Previously, users were involved only in the analysis and design phases, as most of the methodologies are "designed around the needs and capabilities of analysts instead of users" (Dean et al. 1997, p. 186). Nevertheless, these days users should be involved from the beginning to the end as s/he will be able to interact with the system more and to provide more feedback to support effective iteration at each step.

Designers need to select as participants the users who are dealing with the system on a daily basis, not the management and technical personnel. The human aspect has the positive aim of "encouraging the setting and achieving of human objectives as an integral part of the design process" (Mumford 1995, p. 11).

3.2.2   Managing User Participation in Development Processes

Before adopting a participative approach to system development, it is very important to estimate the functions, structures, and processes of participation and to understand the relationship between the management, technical personnel and finally, the more important source, the users. Participation can play a significant role in promoting and endorsing the development process, as participation will "lead to successful outcomes in terms of more information system usage, greater user acceptance, and increased user satisfaction" (Lin and Shao 2000, p. 283). Indeed, "participation is morally right – people should be able to determine their own destinies" (Mumford 1995, p. 13). It enables users to learn more about the system before implementation, producing an "interested and committed group of staff and therefore assisting in the avoidance of morale and job satisfaction problems" (Mumford 1995, p. 13).

Typically, user responsibilities in the participation stage will extend from the beginning until the end of the development process, including the testing and evaluation of the system. For example, user responsibilities can involve "project initiation, determining system objectivities and information needs, identifying sources of information, analyzing information flows, developing input and output formats/screens, and specifying aspects of the user interface" (Doll and Torkzadeh 1989, p. 1155).

Participation is considered a valuable experience for some users who will be involved in the system development process since they will obtain more knowledge, experience about the system before it is implemented. Furthermore, Hartwick and

Barki (1994) indicate that users who participate in the system development process are likely considered that the new system is important and good.

Users will be interested in and attracted to the participation process, as it will:

- **Enable them to** "prevent things that they believe to be undesirable from happening";
- **Avoid and prevent the** "users to undertake tasks that they regard as time-consuming and irrelevant or even being made redundant";
- **Help the users** to make their job more interesting, providing "better services to the client consumers, promotion, and improved quality of working life;"
- **Enhance group harmony,** as it develops a "sense of cooperation and community and produces a willingness to accept group decisions".

(Mumford1995, p. 13)

Although these theories of participation have been primarily developed in the context of design of information systems, they apply equally to the development of websites. Merrick (2001, p. 67) states, "it's important to reach online-users because they are generally the most profitable"

3.2.3   How to Participate?

Participation has a different significance and sense for different groups and individuals, as they have different objectives. Management and designers need to act as a team to present a set of processes and structures that will help the users to achieve their objectives. These gains "will not necessarily be all of the same kind but they should enable each group to say with conviction "participation has clear benefits for us"" (Mumford 1995, p. 13).

The participation process needs to be examined very carefully by both parties (designers and management) to decide which participative approaches should be adopted for the particular development process. There are two main types of participation: indirect "where user representatives participate in the system development process"; and direct "where the users themselves fully participate in the development process" (Barki and Hartwick 1989, p. 54).

Each participation type has special techniques and particular requirements when it is adopted for the development process. For example, if the indirect approach is chosen, then the most important issue that needs to be addressed is to ensure that all interests are represented. Users should decide "how the members of the participative forum are selected or elected and whether a number of groups at different organizational levels are required" (Mumford 1995, p. 14). Whilst, if the direct participation approach is adopted in the development process, the designers and management need to define various issues at the beginning; for example, the degree of participation and the degree of influence that users will have regarding changing aspects of the design, before the implementation.

Users can play a significant role in the development process and this involvement and participation can be in the beginning, middle or at the end of the development process. Each step of this participation has specific requirements and procedures that must be followed so that users can play their role in developing the new system, with anticipation that it will meet their desires and requests.

Mumford (1995) provides a slightly more complex model of participation options. She notes three types of involvement: consultative, representative and consensus. Each one has specific requirements from the users and designers' perspectives.

- **The Consultative approach:** is very useful to secure agreement and settlement between the users and designers at the beginning, to define the objectives of the new system. This approach will allow the full hierarchy of people (top, senior, and low management and interested subordinate staff) to work together to define organizational future needs with respect to the new system. However, "consultative structure must exist or be created so that this sounding out of opinion can be thorough and accurate" (Mumford 1995, p. 18).
- **The Representative approach:** is very appropriate at the definition stage. It is considered useful and powerful since a hierarchy of people will contribute to system definition and setting the boundaries of the new system. A representative approach requires input from all the functions and levels in those parts of the organizations that are using the information system. The design group "will see an important part of its task as involving its departmental colleagues in the design activities and in the decision taking on how work is to be reorganized around the technical system" (Mumford 1995, p. 18).
- **The Consensus approach:** is more popular in most organizations as it enables all the staff associated with developing a new system to take part and have a role in designing the new system for an organization. This is achieved "when efficiency and job satisfaction needs are being diagnosed through feedback and discussion in small groups" (Mumford 1995, p. 18).

It is important to note that each approach has specific time constraints, needs, activities, and potential problems. For example, the consensus approach "does not always emerge easily, and conflicts which result from different interest within a department may have to be resolved first" (Mumford 1995, p. 19). Hence, the other approaches (representative or consultative) are often adopted when developing a new system for an organization.

A participative approach is very useful at all stages, as it will "lead to efficiency gains, the creation of high quality customer care and a good work environment, and more job satisfaction for staff" (Mumford 1995, p. 19). According to Mumford, two types of groups should carry out the stages in the process of systems development (i.e. planning, design, implementation and evaluation):

- **The first group** is responsible for steering the project. The purpose of this group is to provide the link between the different people involved in the project. Moreover, the role and responsibility of this group is to define the "objectives and constraints under which the new system is to be developed" (Mumford 1995, p. 19).

- *The second group* is responsible for defining the system design, to support the function or department where the new system will be implemented and introduced. The role and responsibility of this group is to define the problem, environment, system goals, and (the most important aspect) to identify the impacts of the new systems at each level in the organizational hierarchy.

User participation during the system design will lead the user to understand more about the system firstly, and hence, the system will be more productive and efficient. User participation will "improve the quality of design decisions and resultant applications, improve end-user skills in system utilization, develop user abilities to define their own information requirements, and enhance user commitment to and acceptance of the resultant application" (Doll and Torkzadeh 1989, p. 1152). Moreover, "user satisfaction with a system is a component of job satisfaction, one would anticipate a positive relationship between user involvement and user satisfaction" (Lawrence and Low 1993, p. 196). Participation by users in the development process will provide a more accurate and complete assessment of user "information requirements, avoiding development of unacceptable or unimportant features; improving user understanding of the system and finally will lead to decreased user resistance" (Amoako-Gyampah and White 1993, p. 2).

Rondeau et al. (2002, p. 151) stated that "involving product development managers and manufacturing managers (i.e. end-users) in IS-related activities enables firms to build an IS infrastructure that supports cross-functional decision making". System requirements information can be obtained from the user by using the interview method. This method should be introduced in the development process of web sites to gain more information about the "basic content areas of the site" (Fleming 1998, p. 213). Consequently, to meet the user needs, Fleming (1998) suggests that a three-tiered system of goals-(basic), purpose-(oriented), and topic (or audience) should be considered. The basic goals relate to navigation questions such as "Where am I?" Or "Where can I go?" (Applen 2002, p. 305). Moreover, such design approaches should involve user participation. Effective "communication and positive relationships must be cultivated and planned as any other successful component of project management" (Jiang et al. 2002, p. 20). According to Engler (1996, p. 72), these are the steps, which need to be followed, by designers and management simultaneously during the development process:

- *Identify the correct user:* throughout this step, the designer will define the full range of users and plan for gaining customer input, not just internal user input.
- *Involve the user early and often:*
  - Get the user involved in the development process at all stages (i.e. development, implementation and maintenance);
  - Rules and procedures should be established to motivate the users during the development process;
  - Educate and negotiate with the users regarding their roles and responsibilities – "listen to the users' expectations, what does "involvement" mean to them." (Engler 1996, p. 72);

- Assign a Facilitator who comprehends the required relationship between designers, management and the users. On other words "someone with a foot in both worlds" (Engler 1996, p. 72).

- *Create and maintain a quality relationship:* this step can be achieved by meeting, understanding and listening very carefully to the users.
- *Make improvement easy:* finally, the designer needs to learn the following concepts with respect to the users:

  - Learn the user's language;
  - Proactively solicit the user's opinions;
  - Show the user that his/her opinions make a difference;
  - Make sure there's a demonstrated benefit for user involvement.

3.2.4   Some Problems with the Participative Approach

A participative approach is very practical and valuable to the designer and users simultaneously. It is considered "an important mechanism for improving system quality and ensuring successful system implementation" (Baroudi et al. 1986, p. 232) and "is used to gather local intelligence about particular needs and difficulties at different project sites" (Kawalek and Wood-Harper 2002, p. 18).

However, some system developers believe that a participative approach will create problems for the people who are involved in it, especially to the users. Participation in the system's development process can be seen as "manipulative, will impair labor shedding, will entrench poor practice, can lead to poor design, is not cost-effective, and can be dysfunctional because it can lead to political problems" (Lawrence and Low 1993, p. 195). Hirschheim (1985, p. 295) states that participation can lead "to systems which are not only sub-optimal, but take much longer to develop, and is extremely difficult to operationalize".

According to Mumford (1995), a participative approach can create a few problems for some of the people who are involved in the development process, particularly the users. For example, decrease in trust, conflict over election versus selection of representatives, conflicts of interest, and stress. Key issues include communication and consultation; professional systems designer's role; and finally, the functional or departmental manager role. These problems can occur if the management did not determine the desires and requirements of the people who are involved in the development process, particularly the users.

To prevent and resolve these conflicts, the management needs to address two objectives: (a) firstly, establish good communication mechanisms – for instance, establish a weekly group meeting to provide consultation and commutations skills; and (b) secondly, the management must be in continuous contact with the users to confirm whether or not they are on the correct track with the development process. All problems need to "be recognized, brought out into the open, negotiated and a solution arrived at which largely meets the interest of all parties in the situation"

(Mumford 1995, p. 25). Finally, Olson and Ives (1981) stated that "much of the existing research is poorly grounded in theory or methodologically flawed; as a result, the benefits of user involvement have not been convincingly demonstrated" (Cited in Hirschheim 1985, p. 295).

3.3 How We Know Our Users

This section will discuss the following aspects: defining who the users are in general; user's goals, activities, and environment; their special effects on usability specifications; and the techniques for observation of, and listening to, users.

Users include "those who manage direct users, those who receive products from the system, those who test the system, those who make the purchasing decision, and those who use competitive products" (Preece et al. 2002, p. 171). The different types of users are very important concepts in this research as, through them, the interface can be developed in a way, which meets their needs.

The rationale behind involvement of users in website development is: (1) to reduce time in implementation and testing stages; (2) to familiarize the end-users and client customers with the new system before the implementation; and (3) provide job satisfaction and meet the task effectiveness needs of the end-users and client-customers. A user-centered, task-based approach to system development is required as both User and Task analysis needs must be determined and analyzed very clearly at the beginning of the development process, to prevent any problems with respect to high maintenance costs and user frustration. For example, to make the business booming and prosperous, the supplier needs to answer and meet user requirements regarding services, products, and prices.

3.3.1 User Characteristics

In order to design effectively for users, there are a few user characteristics, which need to be defined for any web project, such as "Learning style, tool preference, physical differences, and cultural differences" (McCracken and Wolfe 2004, p. 38). Unless the system is customizable by the users, then it is the 'average' or, most likely, characteristics of the target user population which need to be considered.

- *Cognitive and Learning Style:* Users will have different cognitive and learning styles. For instance, it is useful to distinguish between the user types "'read then do' people or 'do then read'" people (McCracken and Wolfe 2004, p. 38). In other words, do your users want and expect full instructions before starting, or do your users directly work with the interface without any help and instructions?
- *Interface/Interaction Preferences:* the developer also needs to define user differences with respect to their preferred web interaction techniques (Pull down menu,

Windows ...etc.) and pre-fined mode of interaction with the interface (Mouse or Keyboard). Other questions which need to be asked about the users include:

- What computers, interfaces, and browsers are users currently using?
- Do they always use the same ones or are they familiar with a range of versions?
- Where did they learn these tools? School? On-the-job training? On their own?
- How familiar are they with the tools? How often do they use them? When did they learn?
- Are they familiar with technology that is similar to your intended design? Do they understand frames? Pop-up windows? Search commands?

(McCracken and Wolfe 2004, p. 39)

Besides the above information, the designer needs to learn more about the user's knowledge and background in dealing with the interface; for example, are the "users just starting to use the Internet?" (McCracken and Wolfe 2004, p. 39). If they are novices, it is better to observe them and to assess whether the interface will cause problems and frustration. This experience will help the researcher to find out about problems, which could cause frustration, and how these issues can be resolved before the implementation. Other user classifications relate to:

- **Physical Differences:** The designer needs to gather more information about the typical user, such as age, gender, color blindness, and other physical disabilities.
- **Application Domain Differences:** the designers should also collect more information about the background of their users. For example, if the designer needs to design a website for education, then the vocabulary is different from that used for users from different applications domains – dentists, architects or bankers and so on. According to McCracken and Wolfe (2004, p. 41) "What the 'default'" means to a banker is different from what it means to a programmer. Using the appropriate vocabulary will prevent the user from being forced to ask, "Is this the link I want?" and will empower the user with the conviction, "I want this link."

From all the possible types of user characteristics, a particular set of user classifications (taxonomy) must be selected for a specific website project. For instance, Turk (2001, p. 163) recommends consideration of the following key user characteristics:

- Age
- Culture
- Disabilities
- Education Level
- WWW/IT Experience

The designer should consider these various user characteristics in relations to the design of the website, i.e. the level or particular option for each characteristic – for the average user (and the range) for the target user population. Moreover, more questions need to be asked of the users with respect to visiting a website, for example: the purpose behind visiting this website, how they will work with it, and if they are familiar with this website or ones similar to it. These questions will help the designer to gain more information about the users' knowledge of websites.

3.3.2   Knowledge of User Tasks

This stage in the design process focuses on the purpose behind using the website. For example, if the website is part of a formal work procedure, the designer could expect that the users will be well trained to work with the website. The designer also needs to know if their website-based activities will fit into the workflow of the users' business, and they need to understand "what has been done before the work gets to them, and do they know what happens afterwards" (McCracken and Wolfe 2004, p. 42).

Consequently, designers should understand and recognize two things before they work with the users. Firstly, the designer needs to know the purpose behind visiting the website – is it (for instance) to gain information, shopping or entertainment? Secondly, the designer needs to gain more information about the users' job and the degree of "familiarity they have [with] the basic tools of technology" (McCracken and Wolfe 2004, p. 42).

McCracken and Wolfe (2004) suggest that it is important to understand the users' level of expertise. Users with the lowest level of expertise are termed "Novices." This type of user is "learning a skill for the first time." Novices have a poor understanding of the parts of the website and typical use scenarios. Novices "only recognize a few positions and have not developed any such sequences" (Preece et al. 1994, p. 163). As a result, the purpose of visiting the website is often just to complete a particular task, which they believe will achieve their goals. More advanced users may be classified as follows:

- *Advanced Beginner:* this type of user "is focused simply and exclusively on getting a job done as painlessly and quickly as possible" (Hackos and Redish 1998, p. 82). These people are at the developing stage of expertise and they have knowledge of how to deal with this application and to go through it without any tribulations, especially when the steps are direct and easy to follow. However, these users will be very confused if there are many alternatives to choose from, and if they "encounter difficulties, they have trouble diagnosing or correcting the problem" (McCracken and Wolfe 2004, p. 43).
- *Competent Performer:* these types of users are those "who have learned a sufficient number of tasks that they have formed a sound mental model of the subject matter and the product" (Hackos and Redish 1998, p. 84). These people are willing to learn and study by themselves the principles of how to work with this website. These people may prefer working with the website (or system) via a user manual and documentation to accomplish their goals.
- *Expert:* these users "perform the task automatically without consciously having to think about each move" (Preece et al. 1994, p. 163). These people have the

knowledge to perform a wider range of complex tasks and "suggest solutions to problems" (Preece et al. 2002, p. 346). Experts can develop a "repertoire of sequences of moves" (Preece et al. 1994, p. 163), unlike the novices who are able to utilize only a small set of use scenarios.

Preece et al. (2002) provide a further way of classifying users: the 'Primary users' who are likely "to be frequent hands-on users of the system", while the 'Secondary users' are "occasional users or those who use the system through an intermediary, and 'Tertiary' users are those who are affected by the introduction of the system or who will influence its purchase" (Preece et al. 2002, p. 171).

3.3.3   Recruiting Users

With regard to users, "a representative sample must be involved throughout the design process, from the very beginning" (Cato 2001, p. 41), as they can help the designer not only in one stage but in all the stages. Users need to be selected according to their profile of characteristics and according to the areas, which need to be tested in the interface or website. According to Cato, for "observed testing trails, you need to carry out six individual test sessions with users to obtain meaningful and useful results. Recruit six users for think aloud tests, and twelve for co-participation" (Cato 2001, p. 196). These sessions should be "clearly focused, objective, fast, and cost-effective" (Cato 2001, p. 196). More users can be recruited for website testing by putting messages on appropriate bulletin boards, or via a recruitment agency.

When recruiting users for involvement in participative design, it is best to use real users who are dealing with the interface (i.e. website) very frequently. On the other hand, if real users cannot be recruited, the designer needs to work with "surrogates" such as students from universities and colleges who have an interest in working closely with the interface (i.e. websites) and who are reasonably representative of actual users.

Besides the above, designers need to include:

- Members of the steering committee for the project;
- Members of [the] design team or workshops;
- Reviewers who access the user interface;
- Test users [for] usability tests,
- Test users who exercise the system at delivery time to check that everything works correctly; and
- "Knowledge sources of how task and business procedures are currently carried out" (Lauesen 2005, p. 474).

Preferably, the designer should work very closely with the users to understand why they will use the website and to know exactly how and why particular tasks occur (and in what sequence), the types of problems that are facing the users, and the reasons for these. The designer needs to keep in mind that neither the manager nor the developer will be the type of users working with this website (or system), as both of them are in a different category from the users who are dealing with the website as part of their day-to-day work.

Users who are not in the expert category need support and help (i.e. documentation) from the developer to know how to work with this website (or system) to achieve their goals. Help and support are very important to the users, as via this information, the users can figure out which steps are needed to carry out their task. Therefore, documentation should contain clear, sequential steps in the correct order to allow the users to work efficiently to achieve the target.

3.3.4 Techniques for Observing and Listening to Users

Users are the main source of information for developing an interface such as a website. Therefore, a designer needs to acquire this information to develop and build a website. According to McCracken and Wolfe (2004, p. 44), there are a few golden rules which need to be taken into consideration from the designer's perspective, which include listening to users, "preferably in the context of the place where they will use your website"; and talking to the people who "use your website as part of the work they do on the job and to users who access your website without assistance or interaction with others, at home or work".

In this section, several techniques are discussed that will help the designer to gather more information about the users and their tasks. McCracken and Wolfe (2004, p. 49) states, "Users are in the business of doing their jobs, not explaining how they do their jobs, so simply asking 'How do you do your job?' will not give you the insights you need". Hence, appropriate techniques must be used in order to obtain information from users in an efficient and effective manner. Among the available techniques are: Interviews; Questionnaires; Think Aloud; Talk Right After; Protocol Analysis; Focus Group; and Mailed Surveys. They may be described as follows:

- *Interviews:* Set questions should be asked the users to gain more information about the system. Usually, the interviews occur face to face or via telephone. The purpose behind using this technique is to "gain information about a system and how it is, or will be used" (Bonharme 1996). Generally three types of interview can be used:

  - *Unstructured:* are not directed by a script; data, it is rich but not replicable.
  - *Structured:* are tightly scripted, often like a questionnaire. Replicable but may lack richness.
  - *Semi-structured:* combine features of structured and unstructured interviews and use both closed and open questions. (Preece et al. 2002)

- *Questionnaires:* "Collecting users' subjective opinions about a system can remove unpopular and unusable parts early in the design or after delivery. While interviews provide qualitative data, surveys and questionnaires provide quantitative data which can be statistically analyzed" (Bonharme 1996). Generally, two types of questions can be used – open or closed.

- *Open Questions:* the user is free to provide his/her own answer; however, open questions are difficult to analyze in any rigorous way, or to compare, and can only be viewed as supplementary (Dix et al. 1993, p. 433).
- *Closed Questions:* the user is asked to select an answer from a choice of alternative responses. For example, "there are several rating scales to choose from including, 3-point (yes/no/don't know), ranked order (numbering the options in order of preference), and bi-polar (good/bad)" (Bonharme 1996).

• *Think Aloud:* This technique is very simple and easy to use. It involves asking users to comment on their activities and aspects of the interface while working. This technique was developed by Erikson and Simon for investigating people's problem-solving strategies, and is known as "cooperative evaluation as the user sees himself/herself as a collaborator in the evaluation and not simply as an experimental subject" (Dix et al. 1998, p. 427). This technique requires people "to say out loud everything that they are thinking and trying to do, so that their thought processes are externalized" (Preece et al. 2002, p. 365). The role of the designer is very important as s/he tries to keep the users talking while they are working at their task, whatever that task is, be it simple or difficult. The most important aspect of this technique is to listen very careful to the users discussing the work, their experience, and the environment in which they work. One drawback of this technique is that "thinking aloud" consumes some of the users' cognitive capacity and hence may inhibit their use of the system, biasing the results.

• *Talk Right After:* This technique can be used as an alternative to "Think Aloud" technique as some users cannot speak to the designer while they are working, for example a "travel agent, who is helping someone with questions, can't [cannot] speak to the designer and the customers simultaneously" (McCracken and Wolfe 2004, p. 50). Therefore, to prevent any disruption to the user's performance of the task, the designer can take notes about the tasks and later s/he can discuss it with the user.

• *Protocol Recoding:* There are a number of methods and techniques for recording user actions, for example:

- *Paper and Pencil:* This is a low-technology technique, but a cheap and simple method for collection information from the user. This method "will allow the designer to note interpretations and extraneous events as they occur. However, this method has limitations in obtaining "detailed information as it is limited to the analyst's writing speed" (Dix et al. 1998, p. 428).
- *Audio and Video Recording:* In this technique, the user will be taped during his/her work, and later, the designer will study this tape and take notes of the user's activities. Therefore, this technique is very sensitive and responsive, so the user should be informed in this case, to avoid ethical problems.
- *Computer Logging:* is to get the system "automatically to record user actions at a keystroke level" (Dix et al. 1998, p. 428).

- *Focus Group:* This technique is very common in marketing, political campaigning, and social science research. In this technique, a small number of people (between 5 and 10 users) gather together to discuss a number of prepared questions. A mediator runs the meeting. The most important issue is that actual users should be involved in this step to provide more information and to bring consideration of real problems into the discussion. Normally, the session runs for an hour to an hour and a half.

  - The *advantages* of using this technique are:

    Focus group is low cost and easy to do. In addition, it provides quick results and is easy to scale to gather more data.

  - The *disadvantages* of working with this technique are:

    Facilitator needs to be skillful so that time is not wasted on irrelevant issues. Serious problems can occur if one or two people dominate the entire discussion; therefore, the information will be gathered only from two instead of all the users (Preece et al. 2002). Therefore, an "effective facilitator will attempt to draw everyone into the discussion but will not always be successful" (McCracken and Wolfe 2004, p. 51)

- *Mailed Surveys:* This technique is cheaper for distribution to the users who are dealing with the interface. However, a lot of disadvantages can occur while working with this technique, for example (Fink 2012; Lesser et al. 2011):

  - Takes a lot of skill to write questionnaires that will obtain the information you want;
  - Some groups may interpret the questionnaires in their own way and this will affect the results at the end;
  - Very few people respond to the mailed survey and this will affect the results

- *Web Surveys:* are "powerful tools for maintaining respondent interest in the survey and for encouraging completion of the instrument" (Couper et al. 2001, p. 251). This technique is self-administered and involves computer-to-computer communication over the internet, by asking the users to respond to the survey by clicking on radio buttons and adding additional comments in a specific area within the survey regarding the survey questions. Couper et al. (2001, p. 246) states, "Radio buttons are preferred because this allows mouse-only entry. In addition, radio button version would take less time to complete than the entry box version, given the added burden of typing numbers versus clicking a button". Web surveys are cost savings, speedy, offers greater anonymity, convenience and more sustainable compared with the previous techniques since they are designed and aimed to provide a more dynamic interaction between respondent and questionnaire compared with the paper mail survey. However, online surveys have disadvantages such as technical failures, computer viruses, internet crimes, and hacking into the web-based survey; these aspects can lead to a decrease in the response rate (Dillman 2007; Issa 2013).

- *Field Study:* Field studies are "done in natural settings with the aim of increasing understanding about what users do naturally and how technology impacts them" (Preece et al. 2002, p. 342). Field studies help the designers to identify opportunities for new technology, determine requirements for design, facilitate the introduction of technology, and evaluate technology. Furthermore, field studies get the team "immersed in the environment of their users and allow them to observe critical details for which there is no other way of discovering" (Spool 1997).

The designer must consider carefully the data requirements before an interview (or other data gathering technique) is conducted with the users. The designer needs to address the following issues before the interview:

- Understanding the concepts behind the interface;
- Defining the issues, which need to be clarified from the user such as – tasks, problems, and procedures, which need to be followed to accomplish a specific task.

Throughout the above stages, the designer will gather some information about the interface itself, the tasks, problems, and the steps to accomplish the tasks. If the information does not meet their requirements, then it may be better to apply an alternative information gathering technique before moving to the next step in the methodology.

3.3.5   Internet Marketing and User Responses

There are other ways of determining website users' needs and desires. Internet marketing is a new approach, where customers can define "what information they need, what offering they are interested in, and what price they are willing to pay" (Sheth et al. 2001, p. 6).

According to Hoffman and Novak (1996, p. 51), the Internet is an important focus for marketers for several reasons:

- Consumers and firms are conducting a substantial and rapidly increasing amount of business on the Internet;
- The market prefers the decentralized, many-to-many Web for electronic commerce to the centralized, closed-access environments provided by the online services;
- The World Wide Web represents the broader context within which other hypermedia CMEs (Computer-Mediated Environment) exit;
- The Web provides an efficient channel for advertising, marketing, and even direct distribution of certain goods and information services.

Consequently, Internet marketing is using the Internet and web as a medium to provide information to customers globally. Since it changes rapidly, with new tools being developed to attract more customers to use it, it is important to establish the requirements for interactive marketing. This depends on three issues – "direct communication, individual choice, and friendly technology" (Hanson 2000, p. 95). These address the requirements by learning about each customer's attitudes and behaviors.

In the Internet, several tools can be used by the user to gain more information about specific products or by asking the user to give some feedback about the products. Examples of these tools are user response form, forums, and chat rooms. These tools have two advantages: (1) they encourage the user to provide feedback about the website layout or asking questions about the products in general; (2) they reduce the web master's job by posting all the answers in one place, thereby allowing the users to check the answers from one place.

- *User response form:* this type will allow the user to enter his/her message or checking some fields "can vary from checkbox type responses to the provision of text areas" (Darlington 2005, p. 65). Some systems will be capturing the data from the user response and sending the answer to the user via the e-mail.
- *Forums:* are called 'bulletin boards' or 'newsgroups'; this type of facility provides discussion forums for people with similar interests. For example, "they can also serve as a source of feedback as someone can start a discussion by posting comments about a subject another person may answer, to be followed by other people joining and so on, so a thread of linked messages develops" (Darlington 2005, p. 66).
- *Chat rooms:* are called Internet relay chat (IRC) channels and "allow groups of people to exchange live text messages" (Darlington 2005, p. 67).
- *Blogs:* are called "Web log" or "blogging"; this type of facility has the ability to create an online text diary, "made up of chronological entries that comment on everything from one's everyday life to wine and food to computer problems" (Jessup and Valacich 2008, p. 210). This facility can give an easy method of "publishing web pages which can be described as online journals, diaries or news or events listings" (Chaffey 2007, p. 99).

3.4   Conclusion

This chapter discoursed and studied user participation in the system development process, since it is essential to involve users in the design stage to reduce the gap between users and designers' goals and users and computers on the other.

Currently, there are various types of devices in the market i.e. software applications, mobile and portable devices (e.g. iPads, iPhone) but the majority of these devices are still poorly designed and user satisfaction is inadequate. This chapter presented and addressed user participation significance in the design process by discussing several sections in relation how we know our users, recruiting users and managing user participation in the development processes.

User participation is essential in the sustainable design as well as to improve device acceptance amongst the users, and satisfy their needs. Finally, user participation is vital and fundamental in the system development process along with sustainable design to increase users' acceptance and satisfaction.

References

Amoako-Gyampah K, White KB (1993) User involvement and user satisfaction: an exploratory contingency model. Inf Manag 25(1):1

Applen JD (2002) Tacit knowledge, knowledge management, and active user participation in website navigation. IEEE Trans Prof Commun 45(4):302

Barki H, Hartwick J (1989) Rethinking the concept of user involvement. MIS Q 13(1):53

Baroudi JJ, Olson MH, Ives B (1986) An empirical study of the impact of user involvement on system usage and information satisfaction. Commun ACM 29(5):232–238

Bonharme E (1996) Usability evaluation techniques. http://www.dcs.napier.ac.uk/marble/Usability/Evaluation.html. Accessed 20 Apr 2004

Cato J (2001) User-centered web design. Pearson Education Limited, Harlow/New York

Chaffey D (2007) E-business and e-commerce management. Pearson Education Limited, England

Couper MP, Traugott MW, Lamias MJ (2001) Web survey design and administration. Public Opin Q 65(2):230–253

Darlington K (2005) Effective website development : tools and techniques. Pearson Education Limited, Harlow

Dean DL, Lee JD, Pendergast MO, Hickey AM Jr, JFN (1997) Enabling the effective involvement of multiple users: methods and tools for collaborative software engineering. J Manag Inf Syst 14(3):179

Dillman D (2007) Mail and internet surveys "the tailored design method", 2nd edn. Wiley, Ney York

Dix A, Finlay J, Abowd G, Beale R (1993) Human computer interaction. Pearson Prentice Hall, New York

Dix A, Finlay J, Abowd G, Beale R (1998) Human-computer interaction, 2nd edn. Pearson Education Limited, Hertfordshire

Doll WJ (1987) Encouraging user management participation in systems design. Inf Manag 13(1):25

Doll WJ, Torkzadeh G (1989) A discrepancy model of end-user computing involvement. Manag Sci 35(10):1151

Doll WJ, Torkzadeh G (1991) A congruence construct of user involvement. Decis Sci 22(2):443

Engler N (1996) Bringing in the users. Computerworld 30(48):3

Fink A (2012) How to conduct surveys: a step-by-step guide. Sage Publications, Thousand Oaks

Fleming J (1998) Web navigation: designing the user experience. O'Reilly & Associates, Sebastopol

Hackos JT, Redish JC (1998) User and task analysis for interface design. Wiley Computer Publishing, New York

Hanson W (2000) Principles of internet marketing. South-Western College Publishing – Thomson Learning, Ohio

Hartwick J, Barki H (1994) Explaining the role of user participation in information systems. Manag Sci 40(4):440

Hawk SR, Dos Santos BL (1991) Successful system development: the effect of situational factors on alternate user roles. IEEE Trans Eng Manag 38(4):316

Hepworth JB, Vidgen GA, Griffin E, Woodward AM (1992) The enhancement of information systems through user involvement in system design. Int J Inf Manag 12(2):120

Hirschheim R (1985) User experience with and assessment of participative systems design. MIS Q 9(4):295

Hoffman DL, Novak TP (1996) Marketing in hypermedia computer-mediated environments: conceptual foundations. J Market 60:50–68

Issa T (2013) Online survey: best practice. In: Information systems research and exploring social artifacts: approaches and methodologies. IGI Global, pp 1–19. doi:10.4018/978-1-4666-2491-7.ch001

Jessup L, Valacich J (2008) Information systems today: managing in the digital world. Pearson Education Limited, Upper Saddle River

Jiang JJ, Chen E, Klein G (2002) The importance of building a foundation for user involvement in information system projects. Proj Manag J 33(1):20

Kappelman L (1995) Measuring user involvement: a diffusion of innovation perspective. Data Base Adv Inf Syst 26(2/3):65–86

Kawalek P, Wood-Harper T (2002) The finding of thorns: user participation in enterprise system implementation. Data Base Adv Inf Syst 33(1):13–22

Lauesen S (2005) User interface design "a software engineering perspective". Pearson Education Limited, Harlow

Lawrence M, Low G (1993) Exploring individual user satisfaction within user-led development. MIS Q 17(2):195

Lesser VM, Yang DK, Newton LD (2011) Assessing Hunters' opinions based on a mail and a mixed-mode survey. Hum Dimens Wildl 16(3):164–173. doi:10.1080/10871209.2011.542554

Lin WT, Shao BBM (2000) The relationship between user participation and system success: a simultaneous contingency approach. Inf Manag 37(6):283–295. doi: http://dx.doi.org/10.1016/S0378-7206(99)00055-5

McCracken DD, Wolfe RJ (2004) User-centered website development a human-computer interaction approach. Pearson Education Inc., New Jersey

McKeen JD, Guimaraes T, Wetherbe JC (1994) The relationship between user participation and user satisfaction: an investigation of four contingency factors. MIS Q 18(4):427

Merrick B (2001) Eight steps to achieve e-commerce success. Credit Union Mag 67(2):13

Mumford E (1995) Effective systems design and requirements analysis. Macmillan Press Ltd, Great Britain

Olson M, Ives B (1981) User involvement in system design: an empirical test of alternative approaches. Inf Manag 4(4):183

Preece J, Rogers Y, Benyon D, Holland S, Carey T (1994) Human computer interaction. Addison-Wesley, Wokingham

Preece J, Rogers Y, Sharp H (2002) Interaction design: beyond human-computer interaction. Wiley, New York

Rondeau PJ, Vonderembse MA, Ragu-Nathan TS (2002) Investigating the level of end-user development and involvement among time-based competitors. Decis Sci 33(1):149

Saleem N (1996) An empirical test of the contingency approach to user participation in information systems development. J Manag Inform Syst 13(1):145

Sheridan W (1999) Web design is changing. http://www3.sympatico.ca/cypher/web-design.htm. Accessed 27 May 2003

Sheth JN, Eshghi A, Krishnan BC (2001) Internet marketing. Harcourt, Orlando

Spool JM (1997) Field studies: the best tool to discover user needs. http://www.uiconf.com/8/articles/field_studies_article.html. Accessed 14 Apr 2004

Tait P, Vessey I (1988) The effect of user involvement on system success: a conting. MIS Q 12(1):91

Turk A (2001) Towards contingent usability evaluation of WWW sites. In: Proceedings of OZCHI, Perth, pp 161–167

Chapter 4
Physical, Cognitive and Affective Engineering

Abstract This chapter will assess the importance of physical, cognitive, and affective engineering in designing and developing technology for the user interface, device, and website. These aspects are essential in technology design, since they assist designers to examine the relationships between users and technology, and to improve users' performance when dealing with this technology, in order to reduce errors and increase satisfaction and users' acceptance of the system. These aspects should be part of sustainable technology design to ensure the users' acquiescence, reduce their frustration, and ensure that the new smart technology design will meet user, society, and community needs simultaneously.

4.1 Introduction

In this chapter, the authors explore the physical, cognitive, and affective aspects of engineering. Physical engineering examines how users' physical abilities will interact with and affect the ways in which users perform tasks using technology; cognitive engineering applies knowledge of cognitive attitude in the development of interactive systems. Finally, affective engineering explains how and why users cooperate with technology and how this can be applied to design. This chapter provides to designers and users clear guidelines regarding these concepts, and indicate how and why these concepts are essential in technology design; furthermore, it explains how designers can measure and evaluate physical, cognitive, and affective engineering features in terms of users' requirements.

This chapter is organized as follows: physical engineering, cognitive engineering, GOMS (Goals, Operators, Methods, and Selection rules), Norman's Model, and affective engineering.

4.2 Physical Engineering

This study aims to combine human body mechanics and physical limitations with industrial psychology to facilitate the interaction between human and devices in order to improve people's job performance and cater for users' needs.

© Springer-Verlag London 2015

T. Issa, P. Isaias, *Sustainable Design*, DOI 10.1007/978-1-4471-6753-2_4

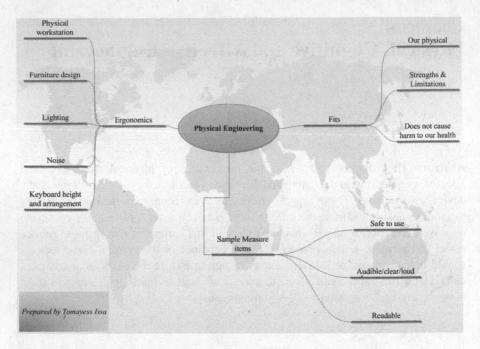

Fig. 4.1 Physical engineering (Prepared by Tomayess Issa)

Physical Engineering aims to improve users' performance ability by handling the work load in the workplace, as improved performance is concerned with reducing errors, improving quality, reducing the time required to complete tasks and ensuring and ascertaining users' acceptance of the system (see Fig. 4.1).

The physical engineering aspects of human computer interaction come into play principally in the process of input and output devices. The main objective of using input devices is to control the system's operations and input data, an example of input devices, mouse, joystick, text, numeric, graphic data, drawing, voice and touch. On the other hand, output devices are machines used to represent data from other devices i.e. monitors, printers, auditory output, synthesized speech, visual display, wearable devices, wireless devices, and haptic devices.

Physical engineering is also concerned with the ergonomics of information systems. It is concerned with things such as the physical workstation and furniture design, lighting, noise, and keyboard height and arrangement. These are all physical aspects of human engineering within an information systems context.

Currently, devices in general are being increasingly used to assist people to improve their job and work performance and productivity. This includes individuals with hearing, vision, or other physical impairment(s). Designers of new smart technologies should consider ways by which to improve the quality of life of people with disabilities, and encourage them to be part of the society and community.

A well-designed computer interface must take into consideration human limitations, since those with disabilities must be considered as members of the community and society in general. Therefore, HCI experts and designers must include these categories of people in their agenda in order to serve them and provide the necessary facilities allowing them to become self-determining and independent. Examples of physical human limitations include (Te'eni et al. 2007; Zhang et al. 2005):

- *Sensory limit*: what and how much our senses can perceive
- *Responder limit*: reach and strength
- *Cognitive limit:* reaction time, accuracy
- *Other limitations:* vision, audition, touch, and motor-related activities

Furthermore, HCI experts and designers should provide the necessary guidelines and principles for accessibility, especially in the new smart technology devices. These guidelines and principles are specified in (Dix et al. 1993; Gerlach and Kuo 1991; Issa and Turk 2010; Te'eni et al. 2007):

- *Standardize Task Sequences:* allow users to perform tasks in the same sequence and manner across similar conditions
- *Ensure the embedded links are descriptive:* using embedded links, the links text should accurately describe the link's destination
- *Use unique and descriptive headings:* use headings that are different from one another and conceptually related to the content they describe.
- *Use radio buttons for mutually exclusive choices:* provide a radio button control when users need to choose one response from a list of equally exclusive options.
- *Non-Text Element:* provide a text equivalent for every non-text element
- *Synchronize:* for any time-based multimedia presentation synchronize equivalent alternatives
- *Color:* information conveyed with color should also be conveyed without it
- *Title:* title each frame to facilitate identification and navigation

Furthermore, Smith and Mosier (1986) offer five high level goals for designing user interface software including the new smart technology and devices for human beings in general:

- *Consistency of data display:* formats, colors, capitalization and so on should all be standardized and controlled by use of a dictionary of these items.
- *Efficient information assimilation by the user*: format should be familiar to the user and should be related to the tasks required to be performed with the data
- *Minimal memory load on the user:* users should not be required to remember information from one screen for use on another screen
- *Compatibility of data display with data entry:* the format of displayed information should be linked clearly to the format of the data entry
- *Flexibility for user control of data display:* users should be able to obtain the information from the display in the form most convenient for the task on which they are working

Furthermore, Shneiderman and Plaisant (2010) establish several guidelines for HCI experts and designers so that their technology design engages users' attention by effectively using features such as intensity, marking, size, fonts, video, blinking, color and audio.

- *Intensity:* use two levels only, with limited use of high intensity to draw attention
- *Marking:* underline the item; enclose it in a box; point to it with an arrow.
- *Size:* use up to four sizes to draw attention
- *Fonts:* use up to three fonts
- *Video:* use opposite coloring
- *Blinking:* use blinking displays or blinking color changes with great care and in limited areas.
- *Color:* use up to four standard colors, with additional colors reserved for occasional use
- *Audio:* use soft tones for regular positive feedback and harsh sounds for rare emergency conditions

Therefore, HCI experts and designers should adopt these guidelines in their agenda and design technologies in order to minimize user frustrations and obstructions and to support disabled people who use devices ranging from workstations to new smart technologies such as iPads or iPhones.

Furthermore, to measure physical engineering, designers must measure safety, audible, and readable. By following these measurements, designers will ensure that the new smart technology meets users' requirements (Shneiderman 1986; Card et al. 1983; Preece et al. 1994).

Finally, several studies (Card et al. 1983; DePaula 2003; Dix et al. 1993; Gerlach and Kuo 1991; Olson and Olson 2003; Preece et al. 1994; Te'eni et al. 2007) indicate that technology and devices are being used more and more to assist users and disabled individuals to accomplish tasks; however, this technology can cause major health risks involving vision and muscular problems, and this can lead to inflammation, disc problems and painful muscles. Therefore, designers should initiate an awareness campaign for the new generation (called internet generation), since these people depend to a great extent on technology for their study and work. This awareness should be available on various media including websites, Facebook and the devices' packaging.

4.3   Cognitive Engineering

Cognitive processes involve user activities including thinking, reading, writing, talking; remembering, making decision, planning, solving problems, and understanding people (see Fig. 4.2). Norman (1993) distinguishes two types of cognition, namely: experiential and reflective. The Experiential mode reflects perceive, act, and react, as it needs a certain level of motivation and enthusiasm, i.e. driving a car, reading a book playing a video game or having a conversation. On the other hand, the reflective mode involves thinking, comparing, and decision-making. This mode leads to creativity and innovation such as writing a book, designing, learning (Isaias and Issa 2015).

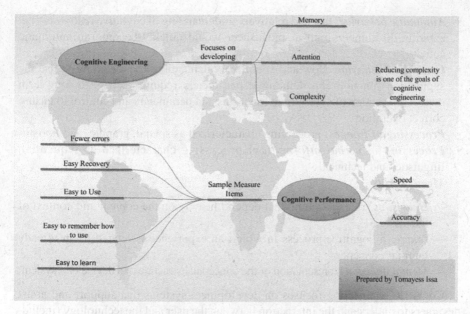

Fig. 4.2 Cognitive engineering (Prepared by Tomayess Issa)

Overall, both modes need specific technologies and are essential for everyday life.

Cognitive Engineering focuses on developing systems which support cognitive processes of users such as memory, perception and recognition, memory, learning, reading, speaking, listening, problem solving, decision making and attention are used in HCI. Difficulty is seen to represent the employment of rare cognitive resources and reducing complication is one of the goals of cognitive engineering (Isaias and Issa 2015).

The human information processing [HIP] model validates how cognitive resources such as memory and processors are employed. There are three types of processors, namely: (1) Perceptual: detects and accepts inputs from the external world and stores parts of the input in the working memory. (2) Cognitive: interprets, manipulates, and makes decisions about the inputs. (3) Motor: is responsible for translating cognitive decisions into physical actions such as using a keyboard. There are two types of memory, namely: working memory which is similar to the human brain's task, since information and data is coming to the human brain for processing and storage of complex cognitive tasks such as language, learning, comprehension and reasoning (Baddeley 1992); and long-term memory which permanently stores, manages and retrieves information for future use and life time (Goelet et al. 1986).

Generally, cognitive engineering takes a narrow view in relation to performance, automatic behavior, controlled behavior, processing of images, processing of verbal information and memory aids (Te'eni et al. 2007, p. 89–90).

• **Performance:** the speed and accuracy of the information-processing task

- *Automatic behavior:* fast and relatively undemanding of cognitive resources (i.e. entering 50 numbers into a spreadsheet would quickly become an automatic activity)
- *Controlled behavior:* slow and cognitively demanding (i.e. deciding to use the summation function and defining it parameters requires access to long-term memory, selection of appropriate functions and parameters and control to ensure correct operation)
- *Processing of Images:* processing characterized as spatial, graphic, and holistic
- *Processing of verbal information:* processing characterized as sequential, linguistic, and technical
- *Memory Aids*

    - *Heuristics:* rules of thumb that depend heavily on the content and context of the task
    - *Image:* a cognitive process in which an experience is related to an already familiar concept
    - *Mental model:* a representation of the conceptual structure of a device or a system

Cognitive engineering focuses on development systems that support and assist designers to understand the interaction between the user and the technology (including computer). Similarly, Gersh et al. (2005) indicate that cognitive engineering developed in response to two reasons, first, to ensure that technologies including computers are well designed and meet users' needs; secondly, it introduced design principles in technology design to ensure that skilled technicians could operate them safely and efficiently.

Finally, in order to measure cognitive engineering, designers should consider the following measures in technology design, namely: fewer errors, easy recovery, easy to use, easy to remember how to use, easy to learn (Dix et al. 1998, 2004).

4.4   GOMS (Goals, Operators, Methods, and Selection Rules)

The GOMS (Goals, Operators, Methods, and Selection rules) model was created by Card et al. (1983). This model aims to present the knowledge of determined human computer interaction (HCI), and how users can interact with computers and the implications for designers. This model endeavours to reduce the complexity in the interface as well as in the cognitive resources and engineering. This model has specific elements that describe purposeful HCI:

- *Goals* specify what the user wants and intend to achieve.
- *Operators* are the building blocks for describing human-computer interaction at the concrete level.
- *Methods* are programs built with operators that are designed to accomplish goals.
- *Selection rules* predict which method will be used. For example, "If the mouse is working, select 'point to an item on screen', if not select 'choose OPEN option in file menu'".

Finally, the GOMS model (Goals, Operators, Methods, and Selection rules) is based on levels of interaction that bridge the gap between the abstract (psychological) task and the concrete (Physical System).

4.5 Norman's Model

To understand the interaction between human and computer, Norman developed a model of user activity (Norman 1986). Before discussing Norman's model, we need to understand the principles of human behavior in order to enhance users' performance in terms of an effective design and technology. These principles are divided into gulf of execution which handles the interruption between the user's goal and aims and its device implementation, and the gulf of evaluation that relates to the gap between device implementation of the user's goal and its evaluation by the user (Te'eni et al. 2007).

Norman's model has eight steps intended to assist users to complete and accomplish a task when using a specific technology:

- *Goals:* create a goal that needs to be accomplished
- *Intentions:* develop an intention that will accomplish the goal
- *Action Specification:* identify a sequence of actions to implement the intentions
- *Execution:* execute the action
- *Perception:* understand the system outcomes from the action
- *Interpretation:* interpret the system state
- *Evaluation:* evaluate the results and compare it with the goals

Figure 4.3 shows the steps that are jointly required the user goals for a particular goal. Generally, these steps will allow users to identify their goals: what is done to the world, the world, and to check the world. In general, these steps have three majors components: identify the goals, do something and evaluate at the end.

4.6 Affective Engineering

Affective engineering focuses mainly on emotions, moods, affective impressions and attitudes; it concentrates on integrating product design and consumers' feelings for a product into design elements (Jordan 2002; Rosson and Carroll 2001; Hewett et al. 1992).

Affective engineering is essential in Human Computer Interaction to balance and integrate the affective and cognitive aspects in the technology design; cognitive engineering interprets and makes sense of the world, while affective engineering evaluates, judges and provides some warning to the users out of possible hazards and risks.

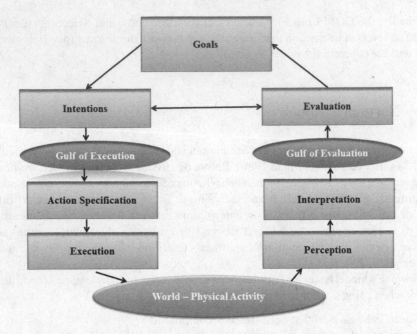

Fig. 4.3 Norman's seven-stage (Adopted from Norman (1986). Prepared by the authors)

Affective engineering is used in any technology design ranging from user interface, technology or websites to color, animation, layout, structure, text, images and menu. For example, using pastel colors for e-commerce sites will leave users feeling calm and will foster a more accepting attitude and readiness to buy and interact further with the site. Additionally, affective engineering focuses on technology design, which is pleasing, engaging, enjoyable, fun, attractive, beautiful satisfying and entertaining. These attributes will encourage the user to accept and use the new smart technology to achieve his/her goals and aims (Fig. 4.4)

Furthermore, user attitudes to combined cognitive and affective engineering are used to evaluate devices including computers, mobiles, and other devices. The evaluation aims to identify errors and problems in order to ascertain whether or not the devices are successful. This is evaluation is based on users' perceptions and opinions and should be taken into account by designers in order to resolve any problems and meet user needs.

Attitudes can be shaped and managed to some extent by training users to examine the devices' performance in general in order to reduce anxiety. Furthermore, a very important step in the design process is the management and involvement of users, as this will promote user satisfaction and acceptance of devices, further reducing user frustration.

Finally, to ensure that users will accept devices, satisfaction is considered the most commonly used in the HCI and information systems field, since users will either confirm or not confirm their satisfaction with the device.

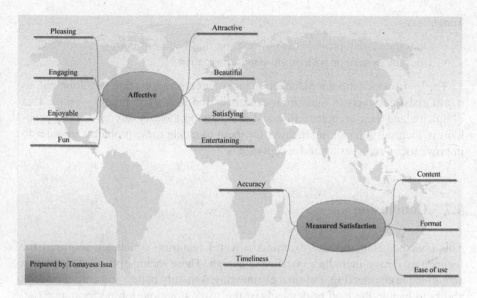

Fig. 4.4 Affective engineering and satisfaction (Prepared by Tomayess Issa)

Doll and Torkzadeh (1988) proposed the most popular measure of satisfaction called End-User Computer Satisfaction. This measure is constructed of five sub-factors namely: content, accuracy, format, timeliness and ease of use.

According to Doll and Torkzadeh (1988, p.268), the five sub-factors include the following aspects:

Content

- Does the system provide the precise information that the user needs?
- Does the information content meet user needs?
- Does the system provide reports that meet user needs?
- Does the system provide adequate information?

Accuracy

- Is the system accurate?
- Is the user satisfied with the system accuracy?

Format

- Is the system output presented in useful format?
- Is the information clear?

Ease to Use

- Is the system user-friendly?
- Is the system easy to use?

Timeliness

- Does the system provide the information that you need in time?
- Does the system provide up-to-date information?

The End-User Computer Satisfaction instrument is a significant development, as it will assist designers to measure user satisfaction with a technology design. This evaluation and measurement will assist designers to identify any errors and problems in their design, making it easier for them to tackle these problems in order to improve users' satisfaction and acceptance.

4.7 Conclusion

This chapter discussed and examined several features, which are required for technology design including sustainable design. These include physical, cognitive and affective engineering. Physical engineering is mainly concerned with the user's ability to handle the load or demands of the work situation, job performance (i.e. reduce errors, enhance quality, and reduce time required to complete specific tasks) and acceptance of the system. Cognitive engineering involves user activities including thinking, reading, writing, talking, remembering, making decision, planning, solving problem and understanding people. This engineering is mainly intended to reduce the complexity between users and devices. Finally, effective engineering works alongside physical and cognitive engineering to examine and assess users' emotions, moods, impressions and attitudes towards product design.

References

Baddeley A (1992) Working memory. Science 255(5044):556–559

Card S, Moran TP, Newell A (1983) The psychology of human computer interaction. Lawrence Erlbaum Associates, Hillsdale

DePaula RA (2003) New era in human computer interaction: the challenges of technology as a social proxy. In: Latin American conference on HCI, 2003, ACM international conference proceeding series, pp 219–222

Dix A, Finlay J, Abowd G, Beale R (1993) Human computer interaction. Pearson Prentice Hall, New York/Harlow

Dix A, Finlay J, Abowd G, Beale R (1998) Human-computer interaction, 2nd edn. Pearson Education Limited, Englewood Cliffs

Dix A, Finlay J, Abowd G, Beale R (2004) Human-computer interaction, 3rd edn. Pearson Education Limited, Harlow

Doll WJ, Torkzadeh G (1988) The measurement of end-user computing satisfaction. MIS Q 12(2):259–274. doi:10.2307/248851

Gerlach JH, Kuo F-Y (1991) Understanding human computer interaction for information systems design. MIS Q 14(4):526–549

Gersh JR, McKneely JA, Remington RW (2005) Cognitive engineering: understanding human interaction with complex systems. J Hopkins APL Tech Dig 26(4):377–382

Goelet P, Castellucci VF, Schacher S, Kandel ER (1986) The long and the short of long-term memory: a molecular framework. Nature 322:419–422

Hewett TT, Baecker R, Card S, Carey T, Gasen J, Mantei M, Perlman G, Strong G, Verplank W (1992) ACM SIGCHI curricula for human-computer interaction. ACM, New York

Isaias P, Issa T (2015) High level models and methodologies for information systems. Springer, New York

Issa T, Turk A (2010) Usability and human computer interaction in developing websites: an Australian perspective. Paper presented at the proceedings of the IADIS international conferences: interfaces and human computer interaction 2010 and game and entertainment technologies 2010, Freiburg

Jordan PW (2002) Designing pleasurable products: an introduction to the new human factors. CRC press, Raton

Norman DA (1986) Seven-Stage model of (individual) interaction. Department of Computer and Systems Sciences, Lulea University, Luleå

Norman D (1993) Things that make us smart. Addison-Wesley, Reading

Olson G, Olson J (2003) Human computer interaction: psychological aspects of the human use of computing. Annu Rev Psychol 54:491–516

Preece J, Rogers Y, Benyon D, Holland S, Carey T (1994) Human computer interaction. Addison-Wesley, Wokingham

Rosson MB, Carroll JM (2001) Usability engineering: scenario-based development of human-computer interaction. Elsevier, Monterey

Shneiderman B (1986) Designing the user interface-strategies for effective human-computer interaction. Pearson Education India, Boston

Shneiderman B, Plaisant C (2010) Designing the user interface: strategies for effective human-computer interaction. Addison Wesley, Boston

Smith SL, Mosier JN (1986) Guidelines for designing user interface software. Mitre Corporation Bedford, Bedford

Te'eni D, Carey J, Zhang P (2007) Human computer interaction: developing effective organizational information systems. Wiley, New York

Zhang P, Carey J, Te'eni D, Tremaine M (2005) Integrating human-computer interaction development into the systems development life cycle: a methodology. Commun Assoc Inf Syst 15:512–543

Chapter 5
Color, Prototyping and Navigation, Principles and Guidelines Design, Evaluation and Testing; Task Analysis

Abstract This chapter discusses the Color, Prototyping and Navigation, Principles and Guidelines Design, Evaluation and Testing, and Task Analysis pertaining to the new smart technology design. These are vital aspects of design that must be taken into account by the designers and HCI experts, by integrating these aspects in the new smart technology design, the new device, user interface, and website will meet the needs of users, the community, and society in general. Therefore, if all these design considerations are taken into account, users will have full control of their devices without any frustration and irritation as users have the opportunity to evaluate and test them in order to meet their needs. Moreover, designers and HCI experts should consider these aspects in their new smart technology design to ensure their new design is in accordance with sustainability principles.

5.1 Introduction

To ensure that new smart technology design is widely accepted and used effectively both globally and locally, designers should consider the following: Color, Prototyping and Navigation, Principles and Guidelines Design, Evaluation and Testing, and Task Analysis. These aspects are essential in any new smart technology design for devices, user interfaces, or websites. Users are becoming more sophisticated and their expectations and behaviors concerning new smart technology design are changing as they have the autonomy to select a new smart technology design, which matches their needs. Therefore, HCI experts should consider the needs of users, the community, and society in order to ensure that the new smart technology design is designed based on sound design principles, which include the notion of sustainability. This chapter is organized as followed: Color, Prototyping and Navigation, Principles and Guidelines Design, Evaluation and Testing; Task Analysis.

5.2 Color

The consideration of Color in a new smart technology design is vital as it can determine the success of failure of a device, interface, or website. Up until now, designers and HCI experts have used color based on their own individual, personal preferences rather than on scientific evidence (Holtze 2006, p. 34).

T. Issa, P. Isaias, *Sustainable Design*, DOI 10.1007/978-1-4471-6753-2_5

This approach will affect users' attitudes to these technologies in terms of style, layout, structure, navigation, usability and ad speed, and their acceptance or rejection of this new smart technology. Shneiderman and Plaisant (2010) posited that designers should limit the number of colors used in their designs, and should select the colors which are the most appropriate for the contents and audience. Furthermore, Te'eni et al. (2007) verified that color usage in new smart technology design will help the user to understand and absorb information when reading, decision making and differentiating between important and unimportant information.

Color theory in new smart technology design is considered in consumer-oriented websites that match the social and emotional perceptions of users, and are expected to "increase trust and be more engaging, also increase user enjoyment or loyalty" (Cyr et al. 2010, p. 2). Color has played an important part in communication, psychology, and even physical health. Arguably, color has power, which is utilized for interior design, graphic design (Web or Interface) and art.

Generally speaking, Color comprises three variables: Hue, Saturation (or Chroma) and Brightness (or intensity or Luminance) (Holtze 2006; Pelet et al. 2013)

- *Hue:*
  - Corresponds to the normal meaning of color – changes in wavelength (these are spectral colors)

- *Saturation (or Chroma)*
  - is the relative amount of pure light that must be mixed with the white light to produce the perceived color

- *Brightness (or Intensity or luminance)*
  - Refers to the shades of Gray decreasing from white through Gray to black

There are three-color wheels, namely Primary Secondary and Tertiary Hues (Morton 2015). Primary Hues: Blue, Red, Yellow (In the printing world these colors are Cyan, Magenta, Yellow); Secondary Hues: Violet, Green, Orange Tertiary Hues: Red-Violet, Yellow-Orange, Blue-Green, Red-Orange, Blue-Violet, Yellow-Green.

The judicious use of color in a new smart technology device has several advantages including: attracting attention, being appealing, facilitating recognition, and assisting memory and comprehension. Moreover, the choice of colors can help users to understand and recall information when undertaking reading and decision-making tasks, and supports effective processes i.e. attract attention, help users to memorize, and add reminders.

There are two general design guidelines for color: firstly, allow for redundancy so that differentiation by color is also accompanied by differentiation by shape or size. Secondly, whenever possible, authorize the users to adapt colors to suit their preferences and their culture.

Let us explain the effects and moods of color usage in new smart technology design. There are various types of colors from cold, cool, hot, warmth, darkness, light, pastel, 'intensity (power, passion) (QSX Software Group 2015; Sibagraphics 2015; Elliot and Maier 2012; Labrecque and Milne 2012)

- **'Cold' colors:**

  - Colors like Blue, green and Blue-green are associated with coldness and calm.
  - Use these colors to promote a feeling of seriousness, significance, honesty, determination, cleanliness, refreshing freshness, coldness.

- **'Cool' colors:**

  - Blue is the base for these colors but added are reds and yellows to bring out a wide range of color from minty green to a soft violet.
  - These colors help promote a feeling of calm, serenity, trust and relaxation.

- **'Hot' colors:**

  - Red is the highest chroma color there is…simply put it is the most powerful hue.
  - A hot color may evoke strong emotional responses, and has been known to stimulate physical activity and sexual desire.
  - Use hot colors if you want an aggressive feel or want something stand out amongst others.
  - Red is the strongest of hues, placing a high chroma yellow in any designed or work of art will draw the eye first.

- **'Warm' colors:**

  - Based in red but softened and suffused with orange and yellows. Warm colors are often used to suggest comfort and warm, heartfelt emotions.

- **'Darkness' colors:**

  - Black is a mysterious color associated with fear and the unknown
  - They are often used to reduce space.
  - These colors are also used so that lighter colors can stand out greater and be more effective.
  - These colors are serious, and can suggest depressed and hardness.

- **'Light' colors:**

  - These colors are barely colors at all; they exist merely as suggestions and hints of colors.
  - They are the opposite of darkness, and they are often used to open up a space or evoke a feeling of openness.

- **'Pastel' colors:**

  - These pale colors are hues tinted with large amounts of white and are very soft in nature.
  - This type of color suggests innocence, fond memories, and romance.

- **'Intensity (Power, Passion)' colors:**

  - The colors of intensity are high chroma colors, pure and seem to scream their message. Great for attention grabbing.

In conclusion, several studies (Wang et al. 2008; Cornforth 1994; Morton 2010) indicate color is essential in new smart technology design as it can enhance marketing, especially in the brand recognition. Compared with black and white, the use of color will increase users' participation and engagement, especially in traditional (i.e. newspapers) and online facilities.

In general, using color in new smart technology design will attract attention, help users to memorize, and add reminders. Moreover, another powerful effect is that it facilitates recognition and comprehension by both the designers and the users.

5.3   Navigation

Navigation is concerned with finding out about, moving through, and the environment. It includes three different but related activities: object identification, which is concerned with understanding and classifying the objects in an environment, exploration, which is concerned with finding out about a local environment and how that environment relates to other environments. Wayfinding, which is concerned with navigation towards a known destination (Elfes 1987; Adler and Blue 1998).

Furthermore, several studies (Blackmon et al. 2002; Fons et al. 2003) indicate that a part of navigation is labelling, as labels are used for internal and external links, headings, subheading, titles, and related areas. For example, there is nothing more confusing for people than a website changing its own vocabulary by referring, for example, to "products" one minute and "items" the next. The same labels should be used consistently on searching mechanisms and on the main pages, in the names of the pages and in the link names.

This type of job will assist the navigation support in any new smart technology design, as many of the signs and labels are deliberately placed in order to support navigation, and it is common to have a navigation bar across the top of a design (i.e. site) which points to the main, top-level categories. This is often called the "global navigation bar".

Within each of these, there will be sub-categories; these might be placed down the left-hand side of the site or may drop down when the main category is selected. This is known as "local navigation".

It is a good design principle to have the same global, top-level navigation bar on every page so that people can easily jump back to the home page, to a "frequently asked questions" page or to one of the other main categories.

An essential aspect of the navigation features of any new smart technology design is to provide a "YOU are here" sign. This is often presented by a description showing where people are in the hierarchy of the site. Other devices such as indexes and glossaries are helpful in assisting people find exactly what they searching for. The site map should be made available so that it can be called up when needed.

One of the significant features of the new smart technology design as an information space is that many sites support the searching process. Search engines can be bought; the better ones are quite expensive but are also effective. Two main problems with searching a website are: the first is knowing exactly what sort of documents the search engine is searching for; the second is how to express a combination of search criteria.

- **Inclusion and Exclusion**
  - With many search engines, you can improve search performance by specifying an *"inclusion operator,"* which is generally a plus (+) sing. This operator states that you do not want a page retrieved unless it contains the specified word. By listing several key terms with this search operator, you can exclude many pages that do not contain one or more of the essential terms. The following, for example, will retrieve only those pages that contain all three of the words mentioned

    i.e. kittens+care+Siamese

- **Wildcards**
  - An asterisk* is a wild card.
  - I.e. Searching for hunt* will return sites with hunter, hunters, hunting, huntsman, etc.

- **Boolean Searches**
  - Use keywords (AND, OR and NOT) to link the words you are searching for.
  - By using Boolean Operators, you can gain a more precise control over your searches.

    i.e. AND operator tells the search service to return only those documents that contain both words

    - i.e. kittens AND care

    i.e. OR operator : is used to search for documents containing either word

    - i.e. Kittens OR care

    i.e. NOT operator tells the search engine to omit any documents containing the word preceded by NOT (just as the minus sign does). For example, the search phrase "kittens NOT cats" retrieves pages that mention kittens but not those that mention cats.

- **Using Parentheses**
  - This operator tells the search engine to search first for what is grouped or nested inside the parentheses.

    i.e. ("kittens" OR "care") AND Siamese

Finally, the basic goals relate to navigation questions such as "Where am I? Or "Where can I go?" (Applen 2002, p. 305). Moreover, such design approaches should involve user participation. Effective "communication and positive relationships must be cultivated and planned as any other successful component of project management" (Jiang et al. 2002, p. 20).

According to Issa (2008), navigation aims to determine the specific navigation paths through the website (including the new smart technology design) between the entities and to establish communication between the interface and navigation in the hypermedia application. Finally, navigation paths are "very important issues to address in website design, for the user has to be able to find what they are looking for as quickly as possible" (Darlington 2005, p. 75). The essential design techniques are: site, layout, link, and navigational structure for the hypermedia application.

5.4 Prototyping

Prototyping is considered a part of the development process and is used to evaluate different proposals for the final website or new smart technology design. Prototyping should be introduced in the new smart technology design (including devices, user interface and website) to identify the layout and the potential problems in the early stages; "functional requirements; navigational issues and visual aspects can also be clarified with the aid of a prototype" (Darlington 2005).

Prototyping can be classified as evolutionary or throw-away. "Evolutionary, means that the prototyping becomes part of the final project", whilst throw-away prototyping "serves only as a pattern for implementation, and you can throw away the prototyping once the interface is complete" (McCracken and Wolfe 2004, p. 8).

Prototyping brings many advantages to the development process that improve communication in the system, including devices, user interface and website, and to remove misunderstanding from requirements in order to demonstrate the object, action or property being discussed, and to provide a basis for an on-going debate with users about their system requirements. Finally, the prototyping approach place(s) greater emphasis on the interpersonal and communication skills of developers and users (Verner and Cerpa 1997).

There are two types of prototyping, namely: low-fidelity and high-fidelity. The latter will be similar to the final product of the website by using software such as Visual Basic, Smalltalk and Macromedia and it is recommended that more than one solution be produced (i.e. three solutions) in order to give the client more options about the 'look' of the website. The advantages of high-fidelity prototyping are: it is very useful for detailed evaluation of the main design elements; it is useful for "selling ideas to people and for testing out technical issues"; (Preece et al. 2002, p. 246). and, it often constitutes a crucial stage in client acceptance – "as a kind of final design document which the client must agree to before the final implementation" (Benyon et al. 2005, p. 254).

Finally, low fidelity prototyping does not look very much like the final product and uses materials that are very different from the intended final version; however, these prototypes are very useful since they tend to be simple, cheap and quick to produce, i.e. storyboarding and sketching (Rudd et al. 1996).

Finally, Issa (2008) confirms that prototyping will allow users and management to interface with a prototype of the new website (including the new smart technology design) to gain some experience in using it. The aims of prototyping are to reduce cost and improve quality during the early stages in the development process.

5.5   Guidelines and Principles Design

To recognize the significance of HCI and Usability features in the web development process as well as in the design process, it is worth scrutinizing the principles and guidelines of design suggested by Te'eni et al. (2007). The implementation of these principles and guidelines when designing and developing a new smart technology, device, user interface, including a website, will improve the presentation, performance, functionality, learnability, efficiency, effectiveness, usefulness or utility; it will reduce errors and inaccuracies in the system, and this will lead to improved user satisfaction and achievement of the goals of both the designer and the user (Leung and Law 2012; Oztekin 2011; Fernandez et al. 2011; Davis and Shipman 2011).

To ensure that the design of a device, user interface, and website will match users' needs, design principles and guidelines are introduced and presented to designers. Principles are used to formalize the high level and widely appropriate design goals while guidelines are essential to the designers to achieve the principles (Zhang et al. 2005; Te'eni et al. 2007). The design principles are divided into seven stages (see Fig. 5.1); each principle focuses mainly on a specific concept, which

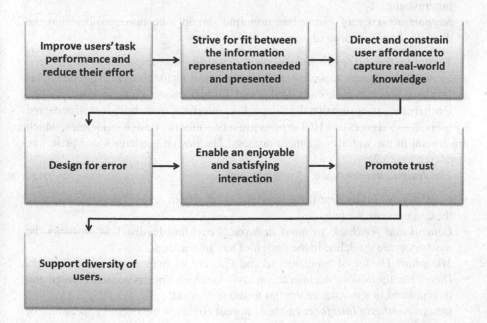

Fig. 5.1 Design principles (Adopted from Te'eni et al. (2007). Prepared by Tomayess Issa)

should be considered from the outset by the designers and users in order to develop a successful device or user interface including a website.

The *design principles* are:

- *Improve users' task performance and reduce their effort:* this principle aims to achieve high functionality along with high usability (i.e. efficiency, ease of use, and comfort in using the system, given that the functionality has been established).
- *Strive for fit between the information representation needed and presented*.

  (a) Representation: a simplified description of a real-world phenomenon.
  (b) Functionality: the set of activities.
  (c) Usability: a measure of ease of use.
  (d) Cognitive fit: system's representation of the problem supports the user's strategies for performing the task.

- *Direct and constrain user affordance to capture real-world knowledge:* the general idea here is that the knowledge required to act effectively resides both in the person's head and in the real world around him/her.
- *Design for error:* a faulty action due to incorrect intention (mistake) or to incorrect or accidental implementation of the intention (slip).
- *Designing for an enjoyable and satisfying interaction:* the design of the interface or website should make the interaction enjoyable for both the designer and the users.
- *Promote trust:* is a critical component in developing an interface or website, especially for the e-commerce systems where the interactions translate directly into revenue.
- *Support diversity of users:* this principle should take into consideration the diversity of populations of users.

To confirm that the device, user interface, or website is widespread and meets users' requirements, designers, especially HCI experts, must include these design principles in their agenda to prevent user frustration and dissatisfaction with these tools.

Furthermore, to ensure that the device, user interface, or website is well accepted by users, the designers and HCI experts must consider the design guidelines, which are crucial in the web development process. The design guidelines comprise five steps (see Fig. 5.2).

The *design guidelines* are:

- *Consistency Guidelines:* If the interface is consistent (even if poorly designed), the end user can adapt to it.
- *Control and feedback go hand in hand:* Providing feedback is probably the most accepted guideline in the design of any interaction.
- *Metaphor:* The use of familiar terms and associations to represent a new concept.
- *Direct Manipulation:* An interaction style in which objects are represented and manipulated in a manner analogous to the real world.
- *Design Aesthetic Interface:* aesthetic appeal concerns the overall appearance of an application.

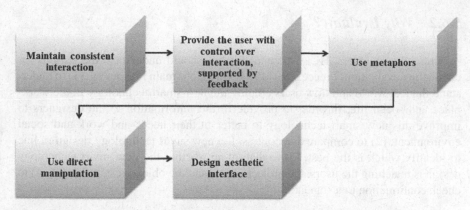

Fig. 5.2 Design guidelines (Adopted from Te'eni et al. (2007). Prepared by Tomayess Issa)

5.6 Evaluation and Testing

This section discusses the importance of the evaluation in the system development process for new smart technology, devices, interfaces and websites. In general, evaluation is an essential step in the system development process, since experts and novices will evaluate the new smart technology, device, interface or website and suggest solutions to problems (Jacobson et al. 1999; Nielsen and Molich 1990).

5.6.1 What is Evaluation?

Evaluation is intended to collect comments and evaluation from the users to ensure that devices, interfaces and websites are meeting the users' needs (Issa 2008). To ensure that the functions of devices, interfaces and websites are effective from the technical perspective, experts and novices test them using specific scenarios. According to McCracken and Wolfe (2004, p. 41), "expert-based evaluation can be achieved by using a group of usability experts to critique the prototype" whilst user-based evaluation can be performed by asking "users to perform representative tasks with the prototype".

Evaluation should occur in the initial stages of the system development process and prior to release to ensure that the device, interface or website matches users' needs. Furthermore, evaluation takes place when the system is released and is used by target users in a real context, that is, during the use and impact stage.

In general, experts and users will evaluate new smart technology, devices, interfaces and websites in terms of usability (i.e. efficient, effective, safe, utility, easy to learn, easy to remember, easy to use, easy to evaluate), HCI (usable, practical, visible, job satisfaction, additional features, text style, fonts, layout, graphics and color) and navigation (site, layout, navigational structure for the hypermedia application) (Issa 2008).

5.6.2   Why Evaluate?

Additionally, designers, HCI experts and users, should understand the reasons for conducting evaluation. Preece et al. (1994) listed four main reasons as: (1) to understand the real world and how users employ the new smart technology in the workplace and social life, in order to provide further information to the designers to improve this new smart technology to better fit their needs and work and social environment; (2) to compare and contrast the new smart technology design in line to identify which is the best; (3) to determine whether the new smart technology design is matching the users, the projects goals and the objectives; and finally 4) to check confirmation to a standard.

5.6.3   When to Evaluate?

In order to ensure that new smart technology design matches users' needs, the designers, HCI experts and users should determine an appropriate time and means of conducting the evaluation. Currently, there are two approaches for formative and summative evaluation. Formative Evaluation: conducted during the development of a product in order to form or influence design decisions. Summative Evaluation: conducted after the product is finished to ensure that it possesses certain qualities, meets certain standards or satisfies certain requirements set by the sponsors or other agencies (Hamilton and Chervany 1981; Nunamaker Jr and Chen 1990; Shackel 1991).

5.6.4   Methods and Means of Evaluation

Real users in real-world contexts can conduct evaluation during the actual use of the produce and this type is called "use and impact evaluation". However, the longitudinal evaluation aims to observe or examine a set of subjects over time with respect to one or more evaluation variables.

To have a successful evaluation, a plan should be formed to identify the stages of design (early, middle, late); the novelty of product (well-defined versus exploratory); number of expected users; criticality of the interface (e.g., life-critical medical system versus museum-exhibit support), costs of product and finances allocated for testing; time available and the experience of the design and evaluation team (Gauthier 2015; Wakefield et al. 2015; Te'eni et al. 2007).

Examples of evaluation strategies include analytical methods (conducted by experts or designers to inspect potential new smart technology design problems), heuristic evaluation (conducted by experts guided by a set of higher-level design principles or heuristics, evaluate to ensure if the new smart technology design is matching the principles and guidelines design). Furthermore, a guidelines review is conducted during the design stage with objective users (i.e. experts or designers outside the design team) to confirm whether the new smart technology design matches the project aims and objectives.

Additionally, cognitive walk-through evaluation is one of the evaluation strategies intended to identify the problems and glitches in the new smart technology design by asking the experts only to evaluate specific tasks in the design; on the other hand, the pluralistic walk-through evaluation will ask experts, designers and users to examine the new smart technology design by considering specific scenarios. This type of evaluation is focused mainly on users' participation and how they would proceed with doing tasks.

In addition, in order to collect from users' further information about the new smart technology design, empirical methods are very useful used i.e. survey/questionnaire, interviews, focus groups, lab experiments, and observing and monitoring usage through field studies. These methods are useful to obtain the necessary feedback from users to improve the new smart technology design and to match users' needs (Nielson and Mack 1994; Shneiderman and Plaisant 2010).

Finally, according to Issa (2008), expert-based and user-based evaluations will test the website to ensure that the web site functions effectively from the technical perspective. Functionality testing and evaluation is mainly about formative usability evaluation by experts and users.

5.7 Task Analysis

To develop a new smart technology which will help to make the devices very successful, the researchers needs to incorporate additional detailed techniques. These will address specific deficiencies identified in the methodologies reviewed in the preceding sections. They relate to:

- detailed task analysis (to facilitate a comprehensive set of links between the front end and back end of an e-commerce websites); and
- detailed procedures for website design and implementation.

It is very important to know one's users when an information system or a website is being developed. At the same time, the designer is required to acquire more information about what users will actually do. To answer this question, the designer needs to adopt a specific technique which is termed 'Task Analysis'. Task analysis is the "process of building a complete description of the [users'] (their) duties" (McCracken and Wolfe 2004, p. 44). This technique involves seeking the following information about the users:

- What tasks they perform
- Why they perform them
- How they perform them

The information will assist designers to determine the basis and foundation for making decisions that will produce successful designs.

5.7.1   Goals, Tasks, and Actions

Participation by users is the basis for developing and creating a simple, easy-to-use user interface or website. Task analysis will help the designer to learn more about the goals and tasks of the users, and in turn to produce an interface that operates effectively and productively.

According to McCracken and Wolfe (2004), goals, tasks and actions should be defined at the beginning of the project. Goals are work-related objectives that include searching for information, sending e-cards, registering a hotel guest, sending e-mail, or doing Internet marketing or non-work related goals such as playing games, chatting or making a plan. Therefore, goals "are technology independent, and they remain the same even when the technology changes" (McCracken and Wolfe 2004, p. 44).

On the other hand, tasks may or may not be consistent between users. Therefore, tasks need to be changed according to the users' requirements and needs, and these tasks are used to accomplish the goals (e.g. buying a book (about HCI) from Amazon.com).

Finally, the last step is action. Actions are "subcomponents of tasks" (McCracken and Wolfe 2004, p. 44). In other words, actions are a series of steps which need to be followed in sequence in order to complete the tasks and, hence, achieve the users' goals. In addition, these steps may involve one or more sub-steps.

5.7.2   Techniques for Identifying Types and Granularity of Tasks

In this section, six techniques will be introduced which can be used to collect more information about the tasks, which are needed to achieve the users' goals. Sometimes, analysts may need to use more than one technique to collect information with respect to the tasks that are needed in order to accomplish the goals.

A key issue is 'Granularity'. This refers to "the level of detail in a description" (McCracken and Wolfe 2004, p. 45). For example, users need to look at their tasks from a short distance to understand its detail as well as from a long distance, to know the purpose behind it. Therefore, in task analysis the granularity that is chosen will depend on "the nature and scope of your website development effort" (McCracken and Wolfe 2004, p. 45).

Workflow Analysis The purpose behind this technique is to illustrate how the work will be done if more than one user is involved in the task. This means that this technique focuses "on work as it passes from person to person" (McCracken and Wolfe 2004, p. 46). As a result, this information may be vast and very helpful for the designer and user simultaneously as it provides a full picture of the project.

Job Analysis This technique is the opposite of the former, as the designer needs to "focus on what a single person does in a day, a week, or a month" (McCracken and Wolfe 2004, p. 46). The designer can collect this information from the users by using the interview method or observing them in their work environment.

Task List This technique takes "the granularity of job analysis to a more detailed level" (McCracken and Wolfe 2004, p. 46). In other words, the designer needs to think very carefully about how many tasks are to be studied in detail before these are broken down into more tasks. In addition, the designer should define and describe the components of a user's job, as some users are responsible for more than one job.

Task Sequences This technique will establish "the order in which the tasks take place" (McCracken and Wolfe 2004, p. 47). The designer can learn the order of these tasks by observing the users at work. However, the important issue which needs to be taken into consideration is to try not to change the users' way of doing the tasks unless there is an important reason for doing so. It is better to give users full control to finalize their job in whatever sequence they like. However, "if you discover that a majority of users do things in a certain sequence, it makes sense to set up the interface to simplify things for the majority" (McCracken and Wolfe 2004, p. 47).

Task Hierarchies The purpose of this technique is to document the components of a task, which are called sub-tasks. The level of detail depends on the type and the purpose of the website.

Procedural Analysis This last technique "contains the most detail of any of the techniques" (McCracken and Wolfe 2004, p. 48). This step will give the designer information about how many steps need to be taken by the user in order to achieve his/her tasks.

Figure 5.3 shows that involving the users in this aspect of the system development process is essential in order to provide the necessary detailed information and to familiarise the users with the new system structure. However, the designer needs to take into consideration the level of user participation in the system development process, which means involving the users in one or more tasks during the process. The user participation level needs to be discussed by the designer and users so that an agreed process can be identified.

Finally, task analysis is essential in the new smart technology development process and involves determining the user types, their work goals and activities, and applies to the device, user interface and website.

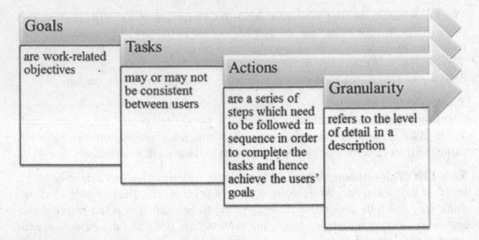

Fig. 5.3 Task analysis (Prepared by Tomayess Issa)

5.8   Conclusion

This chapter has discussed the issues of color, navigation, prototyping, principles and guidelines design, evaluation and testing, and task analysis in terms of new smart technology design. These design concepts are essential especially in new smart technology design, i.e. devices, user interface and website. In general, color is widely used in the development process to attract users' attention and as reminders of specific information on a display. However, navigation enables the user to control the inter-system and intra-system flow of activities and the user's navigation of the system, while prototyping brings designs to life for both designers and users who will use the new design.

Furthermore, this chapter discussed the importance of the evaluation and testing of the new smart technology design as these aspects will assist users and designers to identify the problems and identify some solutions to prevent them in future. On the other hand, to ensure that device, user interface or website is well accepted by designers, and HCI experts must consider the design guidelines, which are crucial in the web development process. Finally, this chapter examined the task analysis focuses on goals, tasks and actions of new smart technology design, and is concerned with logic, cognition, or purpose of tasks.

References

Adler JL, Blue VJ (1998) Toward the design of intelligent traveler information systems. Transp Res C: Emerg Technol 6(3):157–172

Applen JD (2002) Tacit knowledge, knowledge management, and active user participation in website navigation. IEEE Trans Prof Commun 45(4):302

Benyon D, Turner P, Turner S (2005) Designing interactive systems. Pearson Education Limited, Harlow

Blackmon MH, Polson PG, Kitajima M, Lewis C (2002) Cognitive walkthrough for the web. In: Proceedings of the SIGCHI conference on human factors in computing systems, 2002, ACM, pp 463–470

Cornforth D (1994) Color — its basis and importance. In: Pearson AM, Dutson TR (eds) Quality attributes and their measurement in meat, poultry and fish products, vol 9. Advances in meat research. Springer, pp 34–78. doi:10.1007/978-1-4615-2167-9_2

Cyr D, Head M, Larios H (2010) Colour appeal in website design within and across cultures: a multi-method evaluation. Int J Hum Comput Stud 68:1–21

Darlington K (2005) Effective website development. Pearson Education Limited, Harlow

Davis P, Shipman F (2011) Learning usability assessment models for web sites. Paper presented at the IUI 2011, Palo Alto, California

Elfes A (1987) Sonar-based real-world mapping and navigation. IEEE J Robot Autom 3(3):249–265

Elliot A, Maier MA (2012) Color-in-context theory. Adv Exp Soc Psychol 45:61–125

Fernandez A, Insfran E, Abrahão S (2011) Usability evaluation methods for the web: a systematic mapping study. Inf Softw Technol 53(8):789–817. doi:10.1016/j.infsof.2011.02.007

Fons J, Pelechano V, Albert M, Pastor O (2003) Development of web applications from web enhanced conceptual schemas. In: Conceptual modeling-ER 2003, Springer, pp 232–245

Gauthier G (2015) A usability evaluation of a website focusing on the three initial steps of the conflict/resolution process for union members http://scholarspace.manoa.hawaii.edu/bitstream/handle/10125/35855/Gauthier_Final_Paper_Scholarspace.pdf ? sequence=1. Accessed 7 Sept 2015

Hamilton S, Chervany NL (1981) Evaluating information system effectiveness-part I: comparing evaluation approaches. MIS Q 5(3):55–69

Holtze T (2006) The web designer's guide to color research. Internet Ref Serv Q 11(1):87–101

Issa T (2008) Development and evaluation of a methodology for developing websites – Ph D thesis, Curtin University, Western Australia. http://espace.library.curtin.edu.au/R/MTS5B8S4X3B7SBAD5RHCGECEH2FLI5DB94FCFCEALV7UT55BFM-00465?func=results-jump-full&set_entry=000060&set_number=002569&base=GEN01-ERA02

Jacobson I, Booch G, Rumbaugh J, Rumbaugh J, Booch G (1999) The unified software development process, vol 1. Addison-Wesley, Reading

Jiang JJ, Chen E, Klein G (2002) The importance of building a foundation for user involvement in information system projects. Proj Manag J 33(1):20

Labrecque LI, Milne GR (2012) Exciting red and competent blue: the importance of color in marketing. J Acad Mark Sci 40(5):711–727

Leung R, Law R (2012) Human factors in website usability measurement. In: Zhang Y (ed) Future wireless networks and information systems, vol 143. Lecture notes in electrical engineering. Springer, Heidelberg, pp 501–507. doi:10.1007/978-3-642-27323-0_63

McCracken DD, Wolfe RJ (2004) User-centered website development a human-computer interaction approach. Pearson Education Inc., Upper Saddle River

Morton J (2010) Why color matters. http://www.colorcom.com/research/why-color-matters. Accessed 15 May 2015

Morton JL (2015) Basic color theory color matters. http://www.colormatters.com/color-and-design/basic-color-theory. Accessed 1 May 2015

Nielsen J, Molich R (1990) Heuristic evaluation of user interfaces. In: Proceedings of the SIGCHI conference on human factors in computing systems, 1990, ACM, pp 249–256

Nielson J, Mack RL (1994) Usability inspection methods. Wiley, New York

Nunamaker Jr JF, Chen M (1990) Systems development in information systems research. In: System sciences, 1990, Proceedings of the twenty-third annual Hawaii international conference on, 1990, IEEE, pp 631–640

Oztekin A (2011) A decision support system for usability evaluation of web-based information systems. Expert Syst Appl 38(3):2110–2118. doi:10.1016/j.eswa.2010.07.151

Pelet JE, Conway MC, Papadopoulou P, Limayem M (2013) Chromatic scales on our eyes: how user trust in a website can be altered by color via emotion. Adv Intelligent Syst Comput 205:111–121

Preece J, Rogers Y, Benyon D, Holland S, Carey T (1994) Human computer interaction. Addison-Wesley, Reading

Preece J, Rogers Y, Sharp H (2002) Interaction design: beyond human-computer interaction. Wiley, New York

QSX Software Group (2015) Color meaning. http://www.color-wheel-pro.com/color-meaning.html. Accessed 15 May 2015

Rudd J, Stern K, Isensee S (1996) Low vs. high-fidelity prototyping debate. Interactions 3(1):76–85

Shackel B (1991) Usability-context, framework, definition, design and evaluation. In: Brian S, Richardson SJ (eds) Human factors for informatics usability. Cambridge/New York: Cambridge University Press, pp 21–37

Shneiderman B, Plaisant C (2010) Designing the user interface: strategies for effective human-computer interaction. Addison Wesley, Boston

Sibagraphics (2015) The meaning of colours. http://www.sibagraphics.com/colour.php. Accessed 15 May 2015

Te'eni D, Carey J, Zhang P (2007) Human computer interaction: developing effective organizational information systems. Wiley, Hoboken

Verner JM, Cerpa N (1997) Prototyping: does your view of its advantages depend on your job? J Syst Softw 36(1):3–16. doi:http://dx.doi.org/10.1016/0164-1212(95)00193-X

Wakefield B, Pham K, Scherubel M (2015) Usability evaluation of a Web-based symptom monitoring application for heart failure. West J Nurs Res 37(7):922–934

Wang L, Giesen J, McDonnell KT, Zolliker P, Mueller K (2008) Color design for illustrative visualization. IEEE Trans Vis Comput Graph 14(6):1739–1754

Zhang P, Carey J, Te'eni D, Tremaine M (2005) Integrating human-computer interaction development into the systems development life cycle: a methodology. Commun Assoc Inf Syst 15:512–543

Chapter 6
Models and Methodologies

Abstract This chapter examines the various types of models and methodologies
for developing systems (including websites), which may incorporate such HCI processes, usability, and Internet marketing issues. It assesses the advantages and disadvantages of each methodology and analyzes the differences between them in
order to develop the framework for a new participative methodology. To produce a
successful new smart technology, devices "system" (or website), both designers and
users should be working collaboratively. Such user participation has to be facilitated
by a system development methodology consisting of a clear sequence of stages and
steps to be followed by the designer and participating users. The approach of breaking a methodology into stages and steps will be adopted in this research to facilitate
the design process by breaking down the activities into several major stages and
smallest parts into steps (within each stage).

6.1 Introduction

In order for new smart technology, devices, systems, (or websites) to be widely
accepted and used effectively, they need to be well designed. To achieve this, designers and users need to use a specific methodology to produce the "system" (or website). A sound methodology is a very important component of the system
development process, in order to produce a new system, which meets the user's
requirements. A methodology "should tell us what steps to take, in what order and
how to perform those steps but, most importantly, the reasons, 'why' those steps
should be taken, in that particular order" (Jayaratna 1994).

The term "methodology" is used significantly in information systems development,
as each methodology should have a set of stages and steps, which need to be followed
in sequence if the work is to be done successfully. 'Stage' is a "convenient breakdown
of the totality of the information systems life cycle activity" (Olle et al. 1988, p. 21),
while 'step' is "the smallest part of a design process" (Olle et al. 1988, p. 21).

The sequence of the stages may not always be fixed, but it "does suggest that
there is a strict time scale applicable to all situations" (Olle et al. 1988, p. 30). In
some projects, iteration between stages will occur and this may have a range of
impacts on the methodology, as an iteration may "take different forms and thus
impact differently on what one can do with a methodology" (Olle et al. 1988, p. 30).

© Springer-Verlag London 2015
T. Issa, P. Isaias, *Sustainable Design*, DOI 10.1007/978-1-4471-6753-2_6

The main demand is for methodologies that can lead to improvements in the following three aspects according to Avison and Fitzgerald (1993, p. 264): A better end product; A better development process; A standardized process.

For that reason, a designer needs to understand users' requirements for the project before choosing the methodology, in turn to successfully complete the work and to accomplish profitable results.

In this chapter, Issa (2008) will discuss various types of models and methodologies, including: lifecycle models; IS development methodologies; methodologies with explicit human factors aspects; websites methodologies; marketing methodologies; and additional techniques, such as task analysis[1] and detailed website design and implementation. There are numerous similarities in respect to the stages between methodologies for developing information systems, websites, or marketing strategies. Integrating stages from information systems methodologies into website and marketing methodologies is very beneficial in order to develop websites that are more effective and efficient. Human factors experts should be involved in these methodologies to make sure that transaction processes, tracking, maintenance and updating of the website meet the users' requirements.

Firstly, Issa (2008) will discuss the methodologies in this sequence to identify two aspects: (1) the stages needed for the system development process; and (2) the four key principles (user participation, usability, iteration, real interaction), in order to check the availability of these four key principles in IS development, website and marketing methodologies. The system's development cycle will be discussed in order to identify the stages.

Secondly, the stages of information systems development methodologies will be checked to assess how effectively they match the four key principles at each stage and to identify the strongest stage in each methodology. Thirdly, for the website and marketing methodologies, the researcher will: check the availability of techniques covering the four key principles in these methodologies; list the extra stages which will be added to the new methodology; and identify the strongest stage in each methodology.

Finally, additional techniques (i.e. task analysis[2] and detailed website design and implementation) will be discussed. The chapter will also identify any extra stages, which will be added to the new methodology, such as navigation, promotion and staff training. Such additional detailed techniques will play a key role in the new methodology, as most of the existing methodologies have neglected these.

6.2 Lifecycle Models

The term 'lifecycle model' is used to represent a model that captures a set of system development activities and how they are related (Preece et al. 2002). The more sophisticated lifecycle models inform the designer about when and how to move

[1] Task Analysis – Please check Chap. 5
[2] Task Analysis: Please check Chap. 5

from one activity to the next and provide a description of the deliverables for each activity. These lifecycle models are popular since they allow developers, and particularly managers, to get an overall view of the development effort so that processes can be tracked, deliverables specified, resources allocated, targets set and so on. As indicated, some lifecycle models include iteration – this "model incorporates iteration and encourages a user focus" (Preece et al. 2002, p. 186).

The stages in a typical development lifecycle model for interaction design are:

- Define the requirements;
- Prepare some alterative designs, which meet the needs, and requirements that have been identified previously;
- Select a preferred solution;
- Test and evaluate the design;
- Iterate, if necessary. This option can be used either before or after the evaluation stage.

This section discusses and compares a historical sequence of increasingly complex models (i.e. Waterfall Lifecycle Model, Spiral Lifecycle Model, and Rapid Applications Development) which focus on interaction design and adopt the general approach of the development Life Cycle Model.

Furthermore, two models will be discussed in this section from the Human Computer Interaction perspective, the Star Lifecycle Model and Usability Engineering. The former focuses on how the designer addresses Human Computer Interaction design problems, while the latter "shows a more structured approach and hails from the usability engineering tradition" (Preece et al. 2002, p. 192).

6.2.1   The Waterfall Lifecycle Model

This model is basically a linear model where each stage must be completed before the next stage can be started. For example, requirements analysis has to be completed before design can begin. However, iteration can occur at each stage. This lifecycle model is divided into five sequential stages, which may be described as follows:

- **Requirements Analysis:** this stage begins when an organization seeks to add, improve, or correct a system, which is not meeting the requirements of the users. The requirements specification should be captured by the designer in consultation with users to know "what the eventual system will be expected to provide, and how the system will provide the expected services" (Dix et al. 1998, p. 181).
- **Design:** this stage will allow the designers to define the system specifications for the components, such as hardware and software, screen layouts, and documentation.
- **Code:** this stage involves converting design and system specifications into "executable programming language" (Dix et al. 1998, p. 182).
- **Test:** this stage will allow the users to test the new system to ensure that "the system meets their requirements" (Dix et al. 1998, p. 183).

- **Maintenance:** this stage involves the "correction of errors in the system which are discovered after release and the revision of the system services to satisfy requirements that were not realized during previous development" (Dix et al. 1998, p. 183).

One of the main flaws with this model is "that requirements change over time, as businesses and the environment in which they operate change rapidly"; hence, it "does not make sense to freeze requirements for months or years, while the design and implementation are completed" (Preece et al. 2002, p. 188). In addition, although a limited (between stages) iteration option is available in this model, the opportunity to constantly review and evaluate a proposed system with users is not included.

In practice, developing a website by using the waterfall model is complex since most of the users are not "clear how they would want the site to look" (Darlington 2005, p. 34). To solve this problem, prototyping should be introduced since it can help to identify the website layout and the potential problems in the early stages "functional requirements; navigational issues and visual aspects can also be clarified with the aid of a prototype" (Darlington 2005, p. 34).

6.2.2 The Spiral Lifecycle Model

For many years, the Waterfall Lifecycle Model was considered the most popular model for the system development process. However, in 1988 Dr. Barry Boehm introduced the Spiral Lifecycle Model. This model combines the waterfall model with an element called "risk analysis." It is divided into three major stages: (1) planning – to define the objectives, alternatives and constraints; (2) Risk Analysis – for each of the alternatives solutions risks are identified and analyzed; and if this information is not enough, then the prototyping approach will be adopted, before finally, (3) Engineering the solution.

This structured model is very useful as the customer can decide whether any one phase has been completed to his/her satisfaction before the next phase can commence. S/he may elect, if the risks are unacceptably high, to terminate the project. In addition, client evaluation can also be incorporated to check whether or not the system is developing according to their needs.

This model is very useful for large and complex development processes. The regular feedback from the customer allows for any necessary changes to be acted upon immediately. It incorporates steps to identify and controls risks. This model "explicitly encourages alternatives to be considered, and steps in which problems or potential problems are encountered to be re-addressed" (Preece et al. 2002, p. 188). However, if not all aspects of risks are discovered in time, problems will surely occur, thereby leading to the need to repeat the procedures from the beginning, and failure to meet the deadline for accomplishing the project. User involvement is not clearly defined in this model.

6.2.3   Rapid Application Development (RAD)

This model attempts to take a user-centered view and to minimize the risk caused by requirements changing during the course of the project by completing the stages as rapidly as possible. This model has five stages (namely Project set-up; JAD workshops; Iterative design and Build; Engineer and test final prototype; and Implementation Review) and each must be completed before the next stage can be started. However, an iterative approach is incorporated, requiring the developer to go "back to the original data to gather and check the requirements" to determine whether or not it is supporting the user's tasks (Preece et al. 2002, p. 64). RAD added two new key features to the previous development models: Time Boxing and Joint Application Development workshops.

- **Time Boxing** breaks down a large project into many smaller projects. This will allow the designers to deliver the products incrementally and enhances flexibility in terms of the development techniques used and the maintainability of the final system.
- **JAD (Joint Application Development)** workshops between the users and developers are used to gain more information about any difficult issues that are faced and for decisions about system design to be made.

This model also specifically incorporates user testing of prototypes; however, it lacks maintenance of the implemented system. The prototyping in this model should be used to evaluate the system design and to identify the potential problems without any haste. Rapid development and manipulation of a prototype should not "be mistaken for rushed evaluation which might lead to erroneous results and invalidate the only advantage of using a prototype in the first place" (Dix et al. 1998, p. 207).

6.2.4   Systems Development Life Cycle

Kendall proposed the Systems Development Life Cycle in 1992. This lifecycle is a "project management technique that divides complex projects into smaller, more easily managed segments or phases" (FFIEC IT Examination Handbook 2005). The segmentation of projects is a very useful method as it allows the designers and analyst to check if the previous stages have been successfully completed before moving to the next stage. This life cycle is very constructive and useful as it prevents any tribulations to the designer, analysts and users towards the end of the project.

This development life cycle is divided into eight sequential stages (phases), with each needing to be completed before the next stage can be started. The stages are:

- **Initiation Phase:** this stage (phase) begins when an organization decides to add, improve, or correct a system, which is currently not meeting the requirements and needs for the organization and user simultaneously. Consequently, the management needs to define the following requirements before moving to later system development phases:

- Business Considerations (i.e. goals, objectives, budget and legal issues);
- Functional Requirements (i.e. user requirements, hardware and software requirements and backup);
- Project Factors (i.e. project and risk management methodology, and estimated completion dates and costs);
- Cost/Benefits Analysis (including both tangible and intangible benefits and costs).

(FFIEC IT Examination Handbook 2005).

All these requirements need to be considered and support documentation prepared before moving to the planning phase.

- **Planning Phase:** this stage (phase) is very significant as both designers and analysts need to study the requirements very carefully. Throughout this stage, the management needs to address the following items before shifting to the next phase: "communication, defined deliverables, control requirements, risk management, change management, standards, documentation, scheduling, budget, and testing and staff development" (FFIEC IT Examination Handbook 2005).
- **Design Phase:** this stage (phase) allows both the designers and analysts to carry out the design of the new system utilizing the requirements identified by the previous two phases. In this phase, initial prototyping is used to build mock-up designs of items such as applications screens, database layouts, and system architectures. This initial design needs to be reviewed by the users, designers, analysts, network administrators and database managers to make sure it meets the requirements. The initial prototyping design is an iterative process, which means the system will remain in the stage and be reviewed by the participants "until they agree on an acceptable design" (FFIEC IT Examination Handbook 2005).
- **Development Phase:** this stage (phase) involves converting design specifications into an executable program (FFIEC IT Examination Handbook 2005).
- **Testing Phase:** this stage (phase) will allow the users to test the new system to ensure the accuracy of "programmed code, the inclusion of expected functionality and the interoperability of application and other network components" (FFIEC IT Examination Handbook 2005).
- **Implementation Phase:** this stage (phase) will involve installing the new system into the real world environment. In addition, the users' training session for the new system will be carried out.
- **Project Evaluation:** this stage (phase) will allow the management to evaluate and review the "completion of the project objectives and assess project management activities" (FFIEC IT Examination Handbook 2005).
- **Maintenance Phase:** this stage (phase) involves changes and the correction of errors in the hardware, software, and documentation, which are discovered after the implementation stage.

According to L. Peters (1988), this life cycle is a systematic breakdown of the software development process, "... A Software Life Cycle is both a management and a technical tool for organizing, planning, scheduling and controlling the activities associated with a software development and maintenance effort" (cite in Jayaratna 1994, p. 33). However, this life cycle does not allow for significant review and iteration between the stages; this means that suppleness and flexibility for responding to the particular needs of a specific project are missing. It also lacks detailed arrangements for user involvement at all stages.

6.2.5 The Star Lifecycle Model

The Star Lifecycle Model was proposed by Hix and Hartson (1993) to address Human Computer Interaction issues in system development in a more flexible way. This model is six steps namely Implementation, Task/functional analysis, prototyping, requirements specification, conceptual/formal design and evaluation. This model incorporates two different modes of activity: the analytic mode and the synthetic mode. The former is described by concepts such as top-down, organizing, and working from the system view towards the user's view. While the latter is described by concepts such as bottom-up, free thinking, creative and working from the user's view towards the systems view (Preece et al. 2002; Hix and Hartson 1993). The Star Lifecycle Model is extremely flexible and popular, especially with managers, enabling them to get an overview of the "development effort so that process can be tracked, deliverables specified, resources allocated, targets sets and so on" (Preece et al. 2002, p. 193).

The star lifecycle model can be adopted in any system development process and the developer can move from any activity to any other without any specific order as the "activities are highly interconnected" (Preece et al. 2002, p. 193). The evaluation activity is at the center of this model, since, before moving to another activity, one need to pass through the evaluation activity to evaluate the result from the previous activity. This model can be used for defining requirements for a new system, or for evaluating an existing situation and analyzing existing tasks. However, this lifecycle is very general and does not explicitly incorporate procedures for user participation or for system design and maintenance.

6.2.6 The Usability Engineering Lifecycle

Deborah Mayhew proposed the Usability Engineering Lifecycle in 1999, and the purpose of this model is to focus more on how usability design and evaluation tasks may be performed alongside more traditional software engineering activities (Preece 2002).

This lifecycle model presents a "menu of choices that can be worked into the broader development context in order to increase usability" (Instone 2004). It has three main aspects: requirements analysis, design/testing development, and installation. The production of a set of usability goals is the main aspect of the first stage since "these goals [are] captured in a style guide that is [then] used throughout the project to help ensure that the usability goals are adhered to" (Preece et al. 2002, p. 195). The middle stage in this model is the largest and most complex stage as many sub-tasks are involved to produce a detailed design. The final stage involves installation and user feedback.

The most important elements in the Usability Engineering Model are experiential user testing and prototyping, combined with iterative design. "Because it's nearly impossible to design a user interface right the first time, we need to test prototype and plan for modification by using iterative design" (Nielsen 1992, p. 13).

It is anticipated that, via this life cycle, the software engineering discipline "will embrace and incorporate usability engineering and it will become widely institutionalized in development organizations, similarly to how software engineering methodologies in general have become institutionalized" (Mayhew 1999, p. 33). However, this explicitly 'human factors' approach is not easily integrated into the more general technical aspects of other models. This needs to be accomplished by operationalizing the model by using a methodology.

6.2.7 Summary of Lifecycle Models

Several stages were discussed in the lifecycle models section. The stages that are essential for the development of an information system interface, or website, can be summarized as planning, analysis, design, testing, implementation, evaluation, and maintenance. These stages are vital if the designer is to develop an interface, new smart technology or website, which meets the user requirements and needs. However, the models need to be opertationalized as detailed methodologies. As discussed in Chaps. 2 and 3, a critical aspect of systems development is effective HCI; hence, methodologies must adequately address this aspect. Four key principles (user participation, usability, iteration, real interaction) are identified as fundamental aspects in order to develop systems in an effective manner by involving users from the beginning. The four key principles are considered the main foundation for this research to produce websites with high usability, thereby:

- Involving the users in the design from the beginning;
- Avoiding frustrations for the users
- Making the website more approachable, friendly and interesting;
- Winning the trust of the site visitors by meeting users' requirements.

The four key principles are:

- *User participation:* the main purpose is to allow user participation in the website development process to gain more information about the problems, elicit alterative solutions from the users, and familiarize them with the website before it is released;
- *Usability:* to confirm that the website design is efficient, effective, safe, has utility, is easy to learn and easy to remember, usable, practical, provides job satisfac-

tion, and incorporates performance measures that effectively assess the users requirements and requests;

- *Iteration:* to allow for effectiveness and self-correction, this approach will assist the designers to build up the new website and ensure that the project will be tested repeatedly until it meets users' requirements;
- *Real Interaction:* the designer will track users' behavior to present statistics and useful information to demonstrate what attracts or repel users. This can be achieved by adding two options to the web: (1) feedback form to outline users' needs; or (2) adding a counter to a webpage, which will provide detailed statistics (log file) to the designer. Information obtained will include which "Web pages are viewed most often, which domains request Web pages, and what paths users follow as they navigate through a site" (Lazar 2006, p. 44).

In the subsequent sections, the presence of these aspects will be reviewed for each methodology. The rating used for these four key principles will be from 0 to 3. The former presents zero availability while the latter is the maximum. Ratings of 1 and 2 indicate that these aspects are covered in a minimal or moderate way, respectively.

6.3 Information Systems Development Methodologies

System development lifecycle models may be operationalized using methodologies. Information systems development methodologies (ISDM) are an "organized collection of concepts, methods (or techniques), beliefs, values, and normative principles supported by materials resources" (Iivari et al. 2001, p. 186). The main purpose behind using an ISDM is to guide the designer in performing the work by following specific stages in sequence. When developing a system or website, the analyst needs to study the different types of methodologies in respect to their similarities and differences and select the methodology, which best meets the project requirements.

Avison et al. (1993) describe the status of information systems development methodologies as a "methodology jungle". This status of ISDM is "an unorganized collection of numerous methodologies which are more or less similar to each other" (Hirschheim et al. 1998). It was estimated that more than "1000 brand-named methodologies are in use all over the world" (Jayaratna 1994, p. xvii).

It is very difficult for the designer to review the vast array of existing ISDM and check which methodology will accomplish the work to be done. Therefore, the most important aspect of developing a new methodology is "to understand the existing stock and the collective methodology knowledge embedded in them" (Hirschheim et al. 1998). A new methodology should not merely duplicate an existing one but should offer some positive improvement. Consequently, this researcher will develop a new participative methodology for developing websites from the marketing perspective by embedding and grafting stages from various methodologies (Jayaratna 1994) such as those for developing information systems, websites and marketing plans.

Various types of methodologies will be discussed in this section from perspectives of the information systems, human computer interaction, and websites: Structured Systems Analysis and Design Methodology (SSADM); Soft Systems Methodology (SSM); User-Centered Development Methodology; and ETHICS. These methodologies have been chosen for assessment as they cover a range of perspectives, which are likely to address the four key principles identified above.

Such methodologies lay out specific stages to be undertaken and incorporate a range of principles from the lifecycle models discussed in the previous section. This will be presented in a table at the end of each methodology section to address two aspects: (1) checking the level of availability of techniques covering the four key principles in each stage of the methodology; (2) identifying the strongest stage for each methodology. This information will help the researcher in two aspects: (1) to recognize the importance of these four key principles in particular methodologies; and (2) to select stages that will promote the structure of the new participative methodology for developing websites.

6.3.1 Structured Systems Analysis and Design Methodology (SSADM)

This methodology gives the designer "very detailed rules and guidelines to work to" (Avison et al. 1993, p. 191), and "techniques, documentation and training procedures for developing information systems" (Avison and Wood-Harper 1990, p. 181). This methodology is classified into two major parts: three stages of systems analysis and three stages of systems design. The purpose behind this classification is to "make it easier to judge the proportion of time to spend on analysis" (Avison et al. 1993, p. 192). Thus, this methodology is divided into six sequential stages, each of which needs to be completed before the next can be started. The stages are as follows:

- **Analysis of the current system:** investigate and define the problems of the current system.
- **Specification of the required system:** define the aims and services of the new system.
- **User selection of service levels, including technical options:** this stage focuses on users' participation and a feasibility study.
- **Detailed data design:** to define data and the relationships between them, to ensure that the data model meets the requirements of the individual users and the client organization.
- **Detailed procedure design:** this stage is the trial design for the system. The prototype can be paper-based. The user will check if the trial design is working according to their requirements.
- **Physical design control:** develop the system from the paper prototype to an implemented system. The users can use it and test the final system.

Table 6.1 Structured systems analysis and design methodology (Prepared by Tomayess Issa)

Stages	Planning	Analysis	Design	Testing	Implementation	Evaluation	Maintenance
Principles							
User Participation	0	2	1	2	0	0	0
Usability	0	0	0	0	0	0	0
Iteration	0	0	1	0	0	0	0
Real interaction	0	0	0	0	0	0	0
Strongest stage in SSADM	–	–	☑	–	–	–	–

One of the main flaws of this methodology is that it cannot adequately "address the problem of project control and estimating costs directly through the incorporation of project management tools" (Avison et al. 1993, pp. 202–203). In addition, there is limited provision for iteration between stages and maintenance is missing. Table 6.1 indicates that user participation is moderate in the analysis stage.

There is only a minimum rating for user participation and iteration aspects in the design stage to ensure that the data outcomes meet user requirements. Usability and real interaction aspects are rated as zero for each stage of this methodology. The strongest stage in the SSADM methodology is the design stage. This stage will help to identity the data and the relationships between them and produce the trial design for the system. The trial design will be checked by the users to assess if it is working according to users' requirements and requests.

6.3.2 Soft Systems Methodology (SSM)

Checkland proposed the Soft Systems Methodology (SSM) in 1981. SSM provides a "way of tackling messy situations in the real world" (Checkland and Scholes 2003, p. 1). A powerful argument in favor of SSM is that it "has been found to be transferable to people beyond those who developed it, and has been used in several hundred projects around the world" (Rosenhead and Mingers 2002, p. 112). According to Checkland, the SSM methodology involves three roles: client, problem solver, and problem owner. The 'client' "is the person or persons who caused the study to take place" (Checkland and Scholes 1990, p. 47), while, the 'problem solver' "wishes to do something about the situation in question, and the intervention had better be defined in terms of their perceptions, knowledge and readiness to make resources available" (Checkland and Scholes 1990, p. 47). The 'problem solver' is responsible for turning the proposals for change "into real-world action in doing the study" (Checkland and Scholes. 1990, p. 48). The 'problem owner' is the person/group for whom the system has consequences. This methodology is divided into seven sequential stages where each stage must be completed before the next stage can be started.

The stages are as follows:

- **Problem Situation Unstructured (1):** the purpose of this stage is to define the problem and to gain more information and understanding of the problem in general; for example, the SSM should recognize the organization's culture and policies. This can be achieved by meeting the members of the organization and gaining as much information as possible about the organizational structure and culture.

- **Problem Situation Structured (2):** at this stage, the analyst evaluates the problem situation from various approaches and different stakeholders; this means to examine and assess the situation from different worldviews. The stage has several steps: intervention analysis, social and cultural analysis, political analysis, rich picture and utilizes formal and informal methods. The stage has several steps: intervention analysis, social and cultural analysis, political analysis, rich picture and utilizes formal and informal methods.

  - **Intervention Analysis:** this step will help the analyst to define the three roles through which they will learn more about problem situation in general:

    Client: "is the person or persons who caused the study to take place" (Checkland and Scholes 2003, p. 47).
    Problem solver: defines the problem solver, resources and the constraints
    Problem owner: no one is intrinsically a problem owner. The problem solver must decide who is to take [the role of] possible "problem owner" (Checkland and Scholes 2003, p. 47). In addition, the problem owner is the entity "who has a feeling of un ease about a situation, either a sense of mismatch between 'what is' and 'what might be' or a vague feeling that things could be better and who wishes something were done about it" (Checkland 1981, p. 294).

  - **Social and Cultural Analysis:** this step will help the analyst to know more about the internal policies of the organization and to learn more about the motivation and features that effect an individual at the organization. Under this stage, the analyst needs to think about relevant Roles, Norms and Values, as these behaviors nor are fixed, they changed "steadily through time, sometimes slowly sometimes remarkably quickly" (Checkland 1981, p. 231) according to the situation:
  - **Roles:** "a social position recognized as significant by people in the problem situation" (Checkland and Scholes 2003, p. 49)

    Norms: is a "specific prescriptions and proscriptions of standardized practice" (Checkland 1981, p. 231).
    Values: is an "express preferences, priorities or desirable states of affairs" (Checkland 1981, p. 231).

  - **Rich Picture:** is a graphical representation and communication model between the analysts and users to understand system problems and how they can be solved.

- **Formal and Informal Methods:** this step will help the analyst to collect more information about the system by using various methods, informal and formal, such as work observation, interviews and workshops and discussions.

• **Naming of Relevant Systems (3):** this stage aims to involve system-thinking activities. In other words, this stage will involve "formulating of root definitions to a number of relevant systems" (Checkland and Scholes 2003, p. 33). This stage has several steps, such as root definition and CATWOE analysis, which are very important steps as they focus on the human activity systems.

  - **Root Definition:** Checkland and Scholes (2003, p. 33) define root definition as a way to "expresses the core purpose of purposeful activity system". In other words, the core purpose is the transformation process in which some entity 'the input' changes into a new form of entity 'the output'. There are two kinds of Root Definition supported in SSM: 'Primary Task Root Definition' and 'Issue based Root Definition'. The latter is concerned with one-off occurrences (such as a management restructuring), while the former is part of regular activities in the organization.

  - **CATWOE Analysis:** is a way to provide the analyst about with the structure of the real world situation by answering "six element who is doing what for whom, and to whom are they answerable, what assumptions are being made, and in what environment is it happening?" (Avison et al. 1993, p. 247). In other words, "a root definition meeting CATWOE requirements would have driven us more quickly towards aspects which with hindsight we know were finally crucial; we got there in the end, but with CATWOE we should have been quicker" (Checkland 1981, p. 226). According to Checkland et al. (2003, p. 35), CATWOE stands for:

    C: "Customers": the victims or beneficiaries of system activities;
    A: "Actors": people who do the activities;
    T: "Transformation": the conversion of input to output;
    W: "Weltanschauung": the world view which makes this definition meaningful;
    O: "Owners": those who can close the system or stop the event from happening;
    E: "Environment": elements outside the system, which it takes as given

  Two of the major things, which need to be considered, are the T (Transformation) and W (Weltanschauung). The analyst needs to take care with respect to the T (Transformation) as it is "frequently misunderstood, and the systems literature is full of inadequate representations of system inputs and outputs" (Rosenhead and Mingers 2002, p. 74). Moreover, the W (Weltanschauung) might be extreme, such as a "terrorist system" or "freedom-fighting system" (Checkland 1988, p. 244). Therefore, it is essential to declare a "world view when giving an account of any purposeful activity" (Checkland 1988, p. 244).

• **Building the Conceptual Model (4):** this stage is unique and important as it is considered the core of the SSM methodology. It is now required to establish the system requirements from the information, which was gathered from the previous stages. The Conceptual model is a used as "debating point so that the actors can relate the model to the real world situation. Usually there is a conceptual model drawn for each root definition and the drawing up of several root definitions and conceptual models becomes an iterative process of debate and modification towards an agreed root definition and conceptual model" (Avison et al. 1993, p. 247). The stage has several steps: formal system thinking and monitoring the system.

- *Formal System Thinking:* serves as a guideline for checking the conceptual model to determine whether or not it meets the user's requirements.
- *Monitoring the System:* this step will assist the analyst to monitor the system by defining three activities: (1) evaluating the performance in respect to efficacy, efficiency, and effectiveness; (2) monitoring the activities in relation to the problem definition; and (3) taking control action.

- **Comparison (5):** In this stage, the analyst will compare the conceptual models developed in stage four (4) with the definition of the problem situation in stage two (2). The purpose behind this comparison is to define and analyze the differences and similarities between the model and the real world in order to have a "well-structured and coherent debate about a problematical situation in order to decide how to improve it" (Checkland et al. 2003, p. 42).
- **Definition of Desirable and Feasible Changes (6):** this stage is important as the analyst will define those changes that are most feasible and desirable, bearing in mind such considerations as cost and benefit behind the change. It is very important to take into consideration these issues especially before the implementation stage in order to have positive outcomes, which meet the system needs.
- **Recommended Action (7):** this stage defines the changes to the system, and these recommendations should have the approval of the top level in the management before the implementation.

This methodology is a flexible process, as most of the stages can be iterated within the process if improvement is needed. The Soft Systems Methodology seeks to "create a system of enquiry which may be used to examine problem situations and lead to action decisions at both the level of what is required, and how the requirement can be met" (Cropley and Cook 1999, p. 4).

The SSM methodology was created to support the human factors activities in complex existing and new systems. SSM is useful for two reasons: (1) it "bring clarity to confused situation and finding systems solutions in the world of human affairs using 'systems'" (Checkland 2000, pp. 807–813); (2) it helps an organization to allow their systems "less fragmented, less random, more organized, more capable of generating insights and producing commitments" (Checkland 2000, p. 823). This methodology is not appropriate for all situations, as it requires a large gathering of information and often it involves human factors in various stages of the methodology. This methodology is useful when the objectives for the new system need to be clearly defined and clarified and perhaps the most important issue is how the objectives

Table 6.2 Soft Systems Methodology (SSM) (Prepared by Tomayess Issa)

Stages Principles	Planning	Analysis	Design	Testing	Implemen- tation	Evaluation	Mainte- nance
User participation	1	3	2	0	0	0	0
Usability	0	0	0	0	0	0	0
Iteration	1	1	1	1	1	1	1
Real interaction	0	0	0	0	0	0	0
Strongest stage in SSM	☑	☑	☑	–	–	–	–

can be accomplished, via a high-level approach. However, this methodology does not provide for the development of detailed specifications or testing of the system, especially regarding technical aspects. It handles organizational human factors well but does not address detailed design or evaluation of user interfaces.

Table 6.2 indicates that user participation is moderately well utilized in the early stages. Iteration is available in all of the stages with minimum availability to assess if improvement within the system is needed. In contrast, there is a zero rating for usability and real interaction in this methodology. The strongest stages in SSM methodology are planning, analysis, and design. The planning stage examines the nature of the requirements for change and assesses how to address them. The analysis stage will require the analyst to perform the following: (1) evaluate the problem from different angles and from the view of different stakeholders; (2) evaluate the internal policies of the organization; (3) present a graphical presentation (called "rich picture") to the current situation to understand the problem in the system and how to solved it; (4) more informal and formal tools will be used to collect information about the system through. Observation, interviews; workshops and/or discussion. While in the design stage, a small number of considerations should be addressed to identify the purpose behind establishing this system such as: (1) what the system is; (2) how the system will work; and (3) the purpose behind using this system. In addition, users will be involved in the system design and participate in the decision-making.

6.3.3 User: Centered Development Methodology

Another methodology, which may be used to develop successful user interfaces for information systems, is the User-Centered Development Methodology. From the denotation, we learn that this method focuses on involving the user in the process as much as possible, with the ambition that the interface should meet the user's expectation. This can be achieved by user participation within the process activities, such as "observing users while they work, inviting users to participate on the design team and asking users to try out the product and following up on their feedback" (McCracken and Wolfe 2004, p. 5). This methodology involves numerous stages, which focus on "gathering information, designing, building, and testing of a proto-type of the interface" (McCracken and Wolfe 2004, p. 5). It is divided into eight sequential stages, with each needing to be completed before the next stage can be started. The stages may be described as follows:

- **Needs Analysis:** defining the purpose of developing the interface (or website).
- **User and Task Analysis:** defining the users' type and the type of work users will do with the user interface or the website. User and Task analysis focuses on user's goals and their activities, which are carried out by them to achieve their goals. For example, user analysis needs to define: age, education level and user computer knowledge. Task analysis examines user goals. McCracken and Wolfe (2004, p. 7) state that "many products fail because the development team didn't take the time to find out who their users are or what they want to do".
- **Functional Analysis:** defining the functions, which will be available in the inter-face. Through these functions, the users will define their activities in order to achieve their goals.
- **Requirements Analysis:** defining the "formal specifications (i.e. Data Dictionaries, Entity-Relationship Diagrams, and Object-Oriented Modeling) required to implement any system, including websites" (McCracken and Wolfe 2004, p. 7).
- **Setting Usability Specifications:** defining what usability means for the inter-face. For example "performance measure" (i.e. "number of tasks completed", "number of errors" "first impression" and "overall Satisfaction") (McCracken and Wolfe 2004, p. 7).
- **Design:** defining the appearance of the interface, which means, defining the con-tent of the interface and to "organize it according to your user's exceptions". The design "includes the layout of individual pages and how to use visual organiza-tion techniques to create clarity and consistency between pages" (McCracken and Wolfe 2004, p. 7).
- **Prototyping:** developing the initial version of the interface. Prototyping can be classified as evolutionary or throw-away. "Evolutionary, means that the prototyping becomes part of the final project", whilst throw-away prototyping "serves only as a pattern for implementation, and you can throw away the proto-typing once the interface is complete" (McCracken and Wolfe 2004, p. 8).
- **Evaluation:** testing the interface by using expert-based evaluation and/or user–based evaluation. According to McCracken "expert- based evaluation can be achieved by using a group of usability experts to critique the prototype" whilst user-based evaluation can be performed by asking "users to perform representa-tive tasks with the prototype" (McCracken and Wolfe 2004, p. 8). Formative evaluation means "evaluation done during design to check that the product con-tinues to meet users' needs" (Preece et al. 2002, p. 323).

This methodology is "highly iterative and involves as much testing and revision as possible" (McCracken and Wolfe 2004, p. 5). This cycle of repetition can occur in the design, prototype, and evaluation steps, and will be successively run until the interface meets the usability specifications. The most important step is to take into

Table 6.3 User-Centered Development Methodology (UCDM) (Prepared by Tomayess Issa)

Stages Principles	Planning	Analysis	Design	Testing	Implemen- tation	Evaluation	Mainte- nance
User participation	0	1	1	1	0	2	0
Usability	0	0	3	3	0	3	0
Iteration	0	0	1	1	0	1	0
Real interaction	0	0	0	0	0	0	0
Strongest stage in UCDM	–	☑	☑	☑	–	☑	–

consideration user goals and their tasks, as by missing this step, the results will be unsuccessful and unproductive. On the other hand, two basic concepts are missing in this methodology – that is, implementation and maintenance stages. It is also focused on the detail of user interface design without examining the overall relationship between social and technical aspects of the proposed system.

Table 6.3 demonstrates that the four key principles are available in numerous stages with ratings raging from minimum to maximum. User participation is incorporated in analysis, design, testing, and evaluation stages. Testing and evaluation stages are important to ensure that the system meets user requirements. Iteration has minimum rating in design; testing; and evaluation stages. Usability aspects are well covered to ensure user satisfaction with the interface. Finally, the real interaction has zero rating in this methodology.

The strongest stages in the User-Centered Development Methodology are analysis, design, testing, and evaluation. The analysis stage will help the analyst to identify the user's type, goals and the activities, which are carried out by them to achieve their goal. The design stage will define the appearance of the interface. Testing and evaluation stages are included in this methodology, as the interface will be tested by expert-based and user-based evaluation to ensure that the interface or website meets user's requirements.

6.3.4   ETHICS Methodology

Mumford defines a specific methodology with high levels of stakeholder participation called "ETHICS," standing for "Effective Technical and Human Implementation of Computer-based Systems" (Mumford 1995, p. 3). Designers need to involve the user from the beginning, to keep focused on the target audience, to evaluate their activities, and to see if they "address the needs of the contemporary consumer" (Boyer 1999, p. 113). Users, through involvement in the development process, may be able to help to "shape design decisions in ways that deal with their concerns or make their work easier" (Doll and Torkzadeh 1989, p. 1156).

Participation is central to the ETHICS methodology as Mumford defined it as "handing responsibility for the design of a new system to the employees who eventually will have to operate it" (cited in Flynn 1992, p. 300). Two arguments were established from this definition. The first argument is user participation, which needs to be a part of the system development process, whether it be a new or existing system, so that decisions can be made which concern the purpose of the new system. User involvement in the design task can be through groups: "Involvement requires the creation of participative groups, and decisions on the amount and nature of their contribution to the total design process must therefore be made" (Mumford 1995, p. 50).

The second argument is the socio-technical approach that is mainly focused on increasing the ability of the individual to "participate in decision making and in this way to enable him/her to exercise a degree of control over the immediate work environment" (Mumford 1996, p. 70). The members of the Tavistock Institute for two specific reasons created this approach: to create "democratic organizations that are excellent in both human and production terms" (Mumford 1996, p. 73) and to consider the interaction between the social and technical parts of any work system. User involvement in the system development process, according to Mumford, "produces productivity, quality, coordination and control; but also provides a work environment and task structure in which people can achieve personal development and satisfaction" (cited in Flynn 1992, p. 301). Designing and implementing the social–technical approach is not an easy task, as it requires involvement from the users and management simultaneously. Furthermore, this approach requires "training, information, good administration, and skill" (Mumford 1996, p. 77). By adopting these approaches in the new system development process, the outcomes will offer benefits in respect to users' job satisfaction and success of an enterprise.

ETHICS is "pragmatically oriented and relies for its success on the practical abilities and the commitment of the participants to arrive at consensus decisions. It aims to build computer-based information systems which provide job satisfaction and met the efficiency needs of the organization" (Jayaratna 1994, p. 152).

The ETHICS methodology has three objectives focusing on the management of change. These objectives concentrate on the users and their participation in the computer system.

Firstly, the users play a major role in the design of the system, to enrich both job satisfaction and efficiency gains. Mumford said user groups with job satisfaction are able to cope with the required job changes and are "better able to diagnose their own job satisfaction needs than any outside group of specialists" (Mumford 1995, p. 3). An efficiency gain concentrates on user knowledge and the experience in dealing with these interfaces. This experience can be gained by dealing with these interfaces daily, learning about the user needs and system problems. Therefore, users can make a "useful contribution to the specification of the former and the solution of the latter" (Mumford 1995, p. 3).

Secondly, the users are encouraged to contribute to the system design, to define and set satisfaction objectives and to supply additional information to the designer to aid in solving the problems within the system. In addition, the user can contribute his/her experience to explanations of "usual technical and operational objectives" (Mumford 1995, p. 3).

Thirdly, the ETHICS methodology can help ensure that the new technical system is surrounded "by a compatible, well-functioning organizational system" (Mumford 1995, p. 3). This objective is covered by the following concepts:

- Design of work procedures and instructions, for individual work or within groups;
- Define the relationship between the departments or functional areas which the new system will affect;
- The creation of good boundary management techniques;
- Focus on internal and external customers' needs.

(Mumford1995, p. 4).

The ETHICS methodology is basically a linear model where each stage must be completed before the next stage can be started. It involves definition of a set of system characteristics including: why change is needed; systems boundaries; description of the existing system; definition of the key objects and tasks; key information needs; diagnosis of efficiency needs; diagnosis of job satisfaction needs; design of the new system; technical options; preparation of detailed design work; and, implementation and evaluation (Jayaratna 1994).

This methodology recommends many guidelines which are useful for "the understanding and the design of human-centered systems" (Jayaratna 1994, p. 174), and to achieve improvements in efficiency, effectiveness and job satisfaction in the work environment. ETHICS is a "participative design strategy and so employees and users will always be involved in analyzing needs and problem and deciding on solutions" (Mumford 1995, p. 78).

However, the main flaws of this methodology are its inability to handle the "interpersonal and political conflicts that may arise from opening up human feelings and emotions" and its lack of any means, "of discussing or resolving many of the ethical dilemmas that could arise in system development" (Jayaratna 1994, p. 174). In addition, it is quite hard for unskilled users to do the design work appropriately when using this methodology. This methodology does not incorporate iteration between stages, for detailed technical analysis and design or for maintenance.

User participation is dominant in this methodology, to enrich both job satisfaction and efficiency gains. However, there are zero ratings for usability, iteration and real interaction in this methodology. The strongest stage in the ETHICS methodology is the analysis stage. This stage defines the user needs and problems, which allow the analyst to develop a system, which meets the users' requirements and their objectives (Table 6.4).

6.3.5   Summary of Information Systems Development Methodologies

This section will provide a summary behind the Information Systems Development Methodologies

Table 6.4 Ethics methodology (Prepared by Tomayess Issa)

Stages	Planning	Analysis	Design	Testing	Implemen- tation	Evaluation	Mainte- nance
Principles							
User participation	2	3	3	3	3	3	3
Usability	0	0	0	0	0	0	0
Iteration	0	0	0	0	0	0	0
Real interaction	0	0	0	0	0	0	0
Strongest stage in ETHICS	–	☑	–	–	–	–	–

For example, in the **Structured Systems Analysis and Design Methodology (SSADM)** only user participation and iteration stages are available in the design stage, while there is a zero rating for usability and real interaction. The strongest stage in SSADM methodology is the design stage, as this stage will help to define the data and the relationships between them and produce the trial design for the system.

In the **Soft Systems Methodology (SSM)**, numerous techniques for user participation and iteration are available, while there is a zero rating for usability and real interaction. The strongest stages in the SSM methodology are analysis and design. The purpose behind the analysis stage is to evaluate the situation from different angles, and to collect more information to understand the system problem, so as to solve it. The main focus of the design stage is to determine the purpose of establishing this system and involving the user in system design and decision-making.

User-Centered Development Methodology is different from the above methodologies as the four key principles are available in various stages with different ratings, ranging from minimum or maximum availabilities. The most dominant key principle in this methodology is usability to ensure that the interface is easy to learn, easy to use, and with less error frequency, while the real interaction has zero rating in this methodology. The strongest stages in the User-Centered Development Methodology are analysis, design, testing, and evaluation. The analysis will define the type of user(s) and their goals and activities, while the design stage will define the development of the interface. Experts and users combine testing and evaluation phases in one stage to test the interface.

Finally, with the **ETHICS Methodology**, only the user participation aspect is available, to enhance both job satisfaction and efficiency gains, while zero ratings are given for usability, iteration and real interaction. The strongest stage in the ETHICS Methodology is analysis, as via this stage, the analyst will define the users' needs so as to allow the new system to meet their requirements. Table 6.5 illustrates the strongest stages from the Information Systems Development Methodologies analyzed in this chapter and lists the rating availability for the four key principles in each stage. After reviewing the information systems development methodologies and studying each stage, it was noticed that implementation and maintenance were

Table 6.5 Summary of strongest stages in information systems development methodologies (Prepared by Tomayess Issa)

Stage	Information systems development methodologies	Principles			
		User participation	Usability	Iteration	Real interaction
Planning	Soft systems methodology	1	0	2	0
Analysis	Soft Systems Methodology (SSM)	3	0	2	0
	User Centered Development Methodology (UCDM)	1	0	0	0
	Ethics methodology	3	0	0	0
Design	Structured Systems Analysis and Design Methodology (SSADM)	1	0	1	0
	Soft Systems Methodology (SSM)	2	0	2	0
	User Centered Development Methodology (UCDM)	1	1	3	0
Testing	User Centered Development Methodology (UCDM)	1	1	3	0
Implementation	–	–	–	–	–
Evaluation	User Centered Development Methodology (UCDM)	2	1	3	0
Maintenance	–	–	–	–	–

not considered the strongest stages for any of these methodologies, since the focuses of these methodologies are:

- Defining the system problem and clarifying users' needs for the new system;
- Evaluating the current situation and collecting more information to solve the system problem;
- Defining the relationships between the information and produce the trial designs for the system;
- Testing and evaluating the system to ensure that it meets the users' requirements.

However, techniques for effective implementation and maintenance of information systems are included in other (more technical) information system development methodologies not considered above. Since the objective is to develop a methodology for websites, it will be more effective to seek implementation and maintenance techniques targeted to websites. This is addressed in the next section.

6.4   Methodologies for Developing Web Sites

There are many similarities between methodologies for developing information systems and web sites. However, there are also differences. In this section, a range of methodologies from the websites perspective will be discussed in detail, including: Human Factor Methodology for Designing websites; Relationship Management Methodology (RMM); W3DT Design Methodology; Information Development Methodology for the web; and the Web Site Design Method (WSDM). This discussion will define the stages, which need to be carried out, by the designer and users in order to design a website, which meets the user requirements. Most stages focus on feasibility, navigation, deployment, promotion, and measurement of usability and effectiveness.

At the end of each methodology, the researcher will present a table showing: (1) the ratings for the four key principles in each stage within the methodology; (2) the strongest stage for each methodology for developing web sites; and (3) the extra stages available in each methodology. These extra stages will add effectiveness to the new participative methodology for developing websites, and partly reflect the differences between ISDM and website development methodologies.

6.4.1   Human Factors Methodology for Designing Web Sites

Vora (1998) describes a methodology which provides for the development of effective HCI for websites, with the main task being to have a clear understanding of user needs, with particular attention given to: the types of users and their characteristics; and their specific tasks and environments. Vora (1998) also focuses on other important issues in the framework: maintenance, evaluation (expert), and iterative testing (feedback).

This methodology focuses on the human interaction perspective in designing a website. It is basically a linear model where each stage must be completed before the next stage can be started. The stages are as follows:

- **Planning:** the designer needs to answer the following question "Why design a Web Site?" (Vora 1998, p. 155). The stage has several steps: defining the goals; identifying content owners and authors; understanding the users and environments; and finally, the most important aspect is to understand very precisely the users' needs.
- **Analysis:** during this stage, "decisions are made about both content and process" (Vora 1998, p. 156). 'Content' refers to the material necessary to meet identified user tasks, addressing the information needs. The 'process' refers to how the information should be maintained and how "interactive aspects of the websites are handled behind the scenes so that they are transparent to users" (Vora 1998, pp. 156–157).
- **Design and Development:** "information gathered in the earlier stages is translated into actual design" (Vora 1998, p. 160).

- **Usability Testing:** the key to a successful system or (Website) is iterative testing. This testing should occur not only in the final stage, but also in every stage to ensure that the system is on the correct track.
- **Implementation:** this stage is very practical and straightforward, as the designer will transfer the system (or website) to a specific location, to be used by the real user.
- **Maintenance:** this stage is very important. The designer and content providers need to provide up-to-date information on the site to make sure that the changes meet the user needs and to make the site more interesting and useful for the users.

However, this methodology does not specify user participation except in testing and planning. Users can also play a key role in defining content. According to Mayhew, these concepts are very important, especially from the users' perspective, as "One of its great weaknesses, is its lack of quality control for both the content and for presentation" (Mayhew 1998, p. 2). Furthermore, a procedure for addressing user disabilities was missing in Vora's methodology as "designers should keep in mind that the target population includes millions of potential users of Web pages who have various handicapping sensory and physical conditions" (Laux 1998, p. 87). Table 6.6 shows that usability and iteration are the main aspects available in the Human Factor Methodology for Designing Websites. Usability is a very important aspect in this methodology with moderate to maximum rating to ensure that the website meets users' requirements in respect to performance and satisfaction. Iteration is available with minimum and moderate ratings in most stages, to ensure that the system is on the correct track. With respect to user participation, it is available only in the planning, testing and evaluation stages with minimum rating, to identify user goals and understand their environments, and to test the product and make sure it meet users' desires. Finally, the real interaction is available only in the analysis and maintenance stages with moderate to maximum rating to ensure that the website has met users' requirements and – the most important aspect – to make it attractive and approachable to the users.

In the Human Factor Methodology for Designing Websites, there are five strong stages: planning, analysis, testing, evaluation, and maintenance. Planning and

Table 6.6 Human Factor Methodology for Designing Websites (HFMDW) (Prepared by Tomayess Issa)

Stages / Principles	Planning	Analysis	Design	Testing	Implementation	Evaluation	Maintenance	Extra stages
User participation	1	0	0	1	0	1	0	Usability goals development
Usability	2	3	1	3	0	3	0	
Iteration	1	1	1	2	1	2	1	
Real interaction	0	2	0	0	0	0	3	
Strongest stage in HFMDW	☑	☑	–	☑	–	☑	☑	

analysis are essential stages. The former will define the users' goals and examine the environment very carefully in order to meet the users' needs. The main areas of focus of the analysis stage are content (materials to suit user tasks) and process (how information should be maintained). In this methodology, the testing stage is iterative involving "expert evaluation," which means experts will evaluate the website and suggest solutions to problems. Finally, the maintenance stage is also important in this methodology. To make the website more interesting and to attract more users to visit it, designer and content providers need to provide up-to-date information in the site.

6.4.2   Relationship Management Methodology (RMM)

Isakowitz et al. (1995) describe a methodology, which provides for the development of effective websites for highly structured applications such as online conference proceedings, directories, academic journals, courseware and electronic-commerce.

In other words, this methodology is "most suited to applications that have a regular structure, especially where there is a frequent need to update the information to keep the system current" (Isakowitz et al. 1995, p. 43). The main goal of this methodology is to reduce complexity and make the website easy to navigate and maintain, thereby saving time, money, and making it more attractive to the users. This methodology is divided into four sequential stages, where each stage must be completed before the next can be started. The feedback loops between the RMM design stages are shown by dashed lines. While the remaining feedback loops, "although present in RMM, are not shown" (Isakowitz et al. 1995, p. 39).

The stages of RMM may be described as follows:

- **Feasibility:** this stage provides the foundation for the RMM design methodology, as via this stage, the designer needs to define the objectives, user requirements, user analysis, and cost-benefits analysis.
- **Hardware Selection:** this stage involves definition of the hardware requirements for the website.
- **Information/Navigation Requirements Analysis:** during this stage, the designer identifies user tasks and develops an understanding of the information needs and likely use scenarios.
- **Design Methodology:** this stage provides the foundation for designing the relationship between the entities in the web site. The stage has several steps, such as E-R Design, Entity Design, Navigation Design, Conversion Protocol Design, User-Interface Screen Design; and Run-Time Behavior Design.
  - *E-R Design (S1):* this step of the design process "represents a study of the relevant entities and relationships of the application domain" (Isakowitz et al. 1995, p. 39). These entities and relationship of data are considered the basis for the hypermedia applications.

- *Entity Design (S2):* this step is unique to the hypermedia application, as, through it, the designer will determine "how the information in the chosen entities will be presented to users and how they may access it" (Isakowitz et al. 1995, p. 40).
- *Navigation Design (S3):* this step defines how the navigation will be established between the entities, which are based on "associative relationships" (Isakowitz et al. 1995, p. 41).
- *Conversion Protocol Design (S4):* this step sets the conversion rules to "transform each element of the RMDM diagram into an object in the target platform" (Isakowitz et al. 1995, p. 43).
- *User Interface Design (S5):* this step involves the design of screen layouts for each object appearing in the RMDM diagram obtained in Step 3. Via this step, the designer will design the "button layouts, appearance of nodes and indices and location of navigational aids" (Isakowitz et al. 1995, p. 43).
- *Run-Time Behavior Design (S6):* this step considers the "volatility and the size of the domain to decide whether node contents and link endpoints are to be built during application development or dynamically computed on demand at runtime" (Isakowitz et al. 1995, p. 43).
- *Construction and Testing (Evaluation) (S7):* this stage is similar to the one in the traditional software development process. Special care must be taken in this stage to test the website to determine if it is running according to the user requirements, especially the navigational paths.

This methodology is best suited to large websites focusing on product catalogs and hypermedia front-ends of databases. The main flaw of this methodology is that it is missing the maintenance stage. This concept is very valuable, particularly from the users' perception to attract new users to visit the website, and to encourage the current users to visit and work with it. Finally, this methodology does not distinguish "between how information is abstracted and how it is presented. Relationships are just translated to navigational paths and no other communication among the entities is allowed" (Isakowitz et al. 2000). Iteration is available in the design stage with a moderate rating but in the rest of the stages with a minimum rating. The purpose of the iteration stage is to ensure that the website is running according to the user requirements, especially the navigational paths. To prevent any confusion in this methodology, the feedback loops in the design stage were shown as dashed lines, while the remaining feedback present in this methodology is not shown as in the diagram.

There are zero availability ratings for user participation, usability, and real interaction in this methodology. This means that these aspects are not well considered in this methodology.

The strongest stages in the RMM methodology are the planning and design. The planning stage defines the objectives, user requirements and analysis, and cost benefits analysis. While the design stage is the dominant stage in this methodology as the designer will classify: (1) the relationship between the entities in the web site; (2) the navigational path between the entities; and (3) the design of screen and button layouts (Table 6.7).

Table 6.7 Relationship Management Methodology (RMM) (Prepared by Tomayess Issa)

Stages Principles	Planning	Analysis	Design	Testing	Implemen- tation	Evaluation	Mainte- nance	Extra stages
User participation	0	0	0	0	0	0	0	Hardware selection;
Usability	0	0	0	0	0	0	0	navigation design and user interface
Iteration	1	1	2	1	1	1	1	
Real interaction	0	0	0	0	0	0	0	
Strongest stage in RMM	☑	–	☑	–	–	–	–	

6.4.3   The W3DT Design Methodology

Bichler et al. (1996) describe the W3DT (World Wide Web Design Technique), a methodology especially for designing a large-scale Web-based hypermedia application. This methodology focuses on two main parts: modeling techniques and computer-based design. The former gives the designer the possibility to "generate a running prototype of the system, including HTML-pages and CGI-scripts," while the latter allows the designer to define and draw a "graphical representation of a web-site's structure" (Bichler et al. 1996, p. 328). The major requirement for dealing with W3DT is to keep the models "clear and intuitively comprehensible" (Bichler et al. 1996, p. 328).

The essential design primitives and their interaction are best described by the W3DT Meta Model, which shows "the class hierarchy of the different elements" (Bichler et al. 1996, p. 330). The first essential design primitive is Site. One or more diagrams can be found under the site, and each diagram serves two purposes: to indicate a hierarchical refinement of a model; to include sub models into a unified view (Bichler et al. 1996, p. 330).

Usually, a Diagram consists of one page with the option to have "layout" and "link" on the same page. The main purpose of Layout is to hold information about website headers, footers, and background images. On the other hand, the link can be more than just a "hypertext reference to another document" (Bichler et al. 1996, p. 330). Furthermore, page, form, index, and menu are the basic elements for building a "hypermedia application information domain" (Bichler et al. 1996, p. 330). There is no major difference between an Index and a Menu in the W3DT Meta Model, as the former is used to list a complete set of links, while the latter is a "navigational aid with the main purpose to provide access structures" (Bichler et al. 1996, p. 330). It was noted that this methodology has been widely used by several groups of students at universities, colleges, and website developers in organizations "showing very promising results" (Bichler et al. 1996, p. 333). However, this methodology is missing seven essential concepts: planning, analysis, implementation, testing, iteration, evaluation, and maintenance. These stages are very important in the development process as, via them, the designer will test and evaluate the system (or the website) to check whether users' requirements were met.

Table 6.8 The W3DT design methodology (Prepared by Tomayess Issa)

Stages Principles	Planning	Analysis	Design	Testing	Implemen- tation	Evaluation	Mainte- nance	Extra stages
User participation	0	0	0	0	0	0	0	Navigation
Usability	0	0	0	0	0	0	0	design and
Iteration	0	0	0	0	0	0	0	building a
Real interaction	0	0	0	0	0	0	0	hypermedia application
Strongest stage in W3DT	–	–	☑	–	–	–	–	

Table 6.8 indicates zero ratings for the four key principles in the W3DT Design Methodology. This means that none of the above four key principles were incorporated in this methodology to any significant degree. The strongest stage in the W3DT design methodology is the design stage. This stage gives the designer the chance: (1) to generate a first trial product of the system with a hypermedia application; and (2) to draw a graphical representation of the website construction.

6.4.4 Information Development Methodology for the Web

John December (1996) describes a methodology which provides for the development of effective websites for technical communicators, writers, designers and software developers. The main task of this methodology is to decrease difficulty and make the website easy to navigate, maintain, and more attractive to the users. This methodology is very usable for dynamic and competitive web design. December argued that this "methodology was based on the characteristics and qualities of the web on the experiences of web users" (December 1996, p. 372). This methodology is divided into six sequential stages (or elements, according to John December), where each must be completed before the next stage can be started. The stages are as follows:

- **Planning for the Audience and Purpose:** this stage defines several items, which are very useful to build a web site, such as the purpose of the website and audience information. The audience information can include: concerns, background and characteristics. December stated that this planning and analysis requires asking and answering questions such as "Who will use this web? And what will they gain from it?" (December 2003)
- **Setting Objectives and Gathering Domain Information:** after considering the purpose and audience, the designers and analysts need to concentrate on the objectives and goals that the website needs to accomplish.
- **Designing a Web:** to make the web flexible, efficient, and easy to use a relationship should be established between the pages of the web. Therefore, to design a website, the designer should have a thorough grounding in "hypertext, multimedia,

Java and other programming possibilities as well as knowledge about how particular web structures affect an audience" (December 2003).

- **Implementing a Web:** the purpose behind this stage is to create files of HTML and other software. The initial implementation might be a "prototype which is not released publicly, but available for analysis [and use] by a set of representative users" (December 2003).
- **Analyzing a Web:** this stage involves the designer examining the web structure and contents to determine if it meets the objectives, goals, and the purpose of the web.
- **The Web's Release and Promotion and Ongoing Innovation:** involves the web being "publicity released for general web audiences, potential users and current users" (December 1996, p. 372). Furthermore, it involves ongoing support and work to improve the web in order to meet the user requirements.

This methodology is limited to websites for information, art, general services, and entertainment. The methodology is missing two essential aspects: iteration and evaluation stages. These concepts are very important, especially from the users' perspective. Table 6.9 indicated that the four key principles have zero ratings in the Information Development Methodology for the Web except for user participation and real interaction, which have a minimal rating in the implementation stage because of the role of representative users in reviewing the prototype.

The real interaction is available in the maintenance stage to improve the web in order to meet the user needs. The strongest stage in Information Development Methodology for the Web is implementation. This stage releases the first sketch of the website and is checked by representative users in order to make sure it complies with the user requirements.

6.4.5   The Web Site Design Method (WSDM)

Olga De Troyer (1998) describes a methodology for web site design. The main goal for this new methodology is to develop a site which provides information "in such a way that both the provider and the inquirer benefit from it" (De Troyer and Leune 1998, p. 88). The main mission statement for this methodology is [to describe] the

Table 6.9 Information development methodology for the Web

Stages Principles	Planning	Analysis	Design	Testing	Implemen-tation	Evaluation	Mainte-nance	Extra stages
User participation	0	0	0	0	1	0	0	Promotion and prototyping (is available under the implementation Phase)
Usability	0	0	0	0	0	0	0	
Iteration	0	0	0	0	0	0	0	
Real interaction	0	0	0	0	0	0	1	
Strongest stage in IDMW	–	–	–	–	☑	–	–	

subject purpose and the target audience for this website. Without giving good consideration to the mission statement there "is no proper basis for decision making or
for the evaluation of the effectiveness of the website" (De Troyer 1998, p. 53).

This methodology has adopted the "user-centered" approach in order to create
effective communication and to define the different types of users and characteristics and their information requirements. This will lead to definition of the "perspectives." A perspective "is a kind of user subclass", which means, "all users in a user
class with the same characteristics and usability requirements" (De Troyer 1998,
pp. 54–55). This methodology consists of the following stages: User Modeling,
Conceptual Design, Implementation Design and the actual Implementation.

- **User Modeling:** this stage is divided into two steps: User Classification and User
  Class Description. The purpose behind this stage is to concentrate "on the potential users of the Web site" (De Troyer et al. 1998, p. 88).

  - *User Classification:* this step will help the designers to identify the future
    users or visitors of the website and classify them into user classes. Therefore,
    the purpose of this step is to identify the target audience by "looking at the
    organization or the business process which the website should support" (De
    Troyer 1998, p. 53).
  - *User Class Description:* this step will help the designer to analyze in more
    detail the user types in order to identify not only their "information require-
    ments but also their usability requirements and characteristics" (De Troyer
    1998, p. 54). Examples of information requirements are "levels of experience
    with websites in general, language issues, education/intellectual abilities,
    age." Some of this information can be "translated into usability requirements"
    (De Troyer 1998, p. 54).

- **User Conceptual Design:** this stage is divided into two steps: User Modeling
  and the Navigational Design. This stage utilizes different "user classes and their
  perspectives" which will allow the users to efficiently "navigate through the Web
  site" as each user class has its own "navigation track" (De Troyer et al. 1998,
  p. 90).

  - *Object Modeling:* this step will help the designers to identify information
    requirements of different user classes and their perspective.
  - *Navigational Design*: this defines the specific navigation path through the
    website for each user class.

- **The Implementation Design:** this stage will help the designer to design the
  "look and feel" of the website, to "create a consistent, pleasing and efficient look
  and feel for the conceptual design made in the previous phase" (De Troyer 1998,
  p. 55).
- **The Implementation:** is the "actual realization of the website using the chosen
  implementation environment, e.g. HTML" (De Troyer 1998, p. 55).

The WSDM methodology is "user centered" rather than "data driven", which
means the starting, point for this methodology "is the set of potential visitors of the

Web site" (De Troyer et al. 1998, p. 85). The user participation is not strong in this methodology; however, the WSDM methodology seeks to learn more information about the users in respect to their knowledge in dealing with the website, language, education, and age. This information will help the designer to translate these user characteristics into usability needs and requirements of the website. However, the WSDM methodology is missing a few stages in the development process, namely: testing, iteration, evaluation, and maintenance. These stages are important, as, through them, the designer will learn if the website meets users' requirements.

Table 6.10 indicates that user participation is covered in the planning; analysis and design stages with minimal rating, as the designer is seeking to gain more general information about the users such as language, age and education, as some of this information will be translated into usability requirements. Usability aspects are available in planning, analysis, design, and implementation with a moderate rating, while the real interaction has a similar rating but in analysis and design. For iteration, the rating is zero, which means it is not considered in this methodology. The strongest stages in the WSDM are the planning, analysis and design. The planning stage will help the designer to identify the target audience to the website and to classify them into user classes; while the analysis stage will help the designer to analyze in more detail the user types in order to identify information and usability requirements and characteristics. Finally, the design stage will help the designers to identify the information required, how it will be presented, and the navigation paths for user types.

6.4.6 Summary of Methodologies for Developing Web Sites

This section will provide a summary behind the methodologies for developing Web sites:

For example, in the **Human Factor Methodology for Designing Websites,** the four key principles are available but in varying degrees in different stages. Usability is very dominant in analysis, testing and evaluation stages with maximum rating, while in the planning and design stages it has a moderate rating. This means that usability is a very significant aspect in this methodology to ensure that the website

Table 6.10 The Web Site Design Method (WSDM) (Prepared by Tomayess Issa)

Stages Principles	Planning	Analysis	Design	Testing	Implemen- tation	Evaluation	Mainte- nance	Extra stages
User participation	1	1	1	0	0	0	0	User modeling
Usability	2	2	2	0	1	0	0	and
Iteration	0	0	0	0	0	0	0	conceptual
Real interaction	0	2	2	0	0	0	0	design
Strongest stage in WSDM	☑	☑	☑	–	–	–	–	

is running without any errors and enhancing job satisfaction. Iteration is available in some stages with minimum rating that is in planning, analysis, design, implementation, and maintenance, with a moderate rating in testing. User participation is available only in the planning, testing and evaluation stages with a minimum rating, while the real interaction has a moderate rating in analysis, and maximum rating in the maintenance stage. In the Human Factor Methodology for Designing Websites, there are five strongest stages: planning, analysis, testing, evaluation, and maintenance. Planning and analysis are essential stages for defining the users' goals, understanding the environment, and the way that information should be maintained. The testing and evaluation stages are also very important. Finally, the maintenance stage incorporates the provision of up-to-date information, in order to make the website more attractive and interesting.

In the **Relationship Management Methodology (RMM)**, only iteration is available with minimum or moderate ratings in all the stages. Zero rating for user participation usability and real interaction in this methodology means that usability, user participation, and real interaction are largely ignored. The strongest stages in the RMM methodology are design and planning. Design and planning are essential, as the former will help the designer to define the relationship and navigational path between the entities and to design the screen and button layouts; whilst the latter will define users' goals and an understanding of the cost benefits analysis.

The four key principles have zero ratings in **The W3DT Design Methodology** and the **Information Development Methodology for the Web** except for a minimum rating for user participation in the implementation stage and with minimum rating for real interaction in the maintenance of the latter methodology. This means that the four key principles are largely ignored in these methodologies. The strongest stage in the **W3DT Design Methodology** is the design stage. The strongest stage in the **Information Development Methodology for the Web** is implementation. This stage permits the users to check the first draft of the website to ensure it meets the users' requirements and needs.

Finally, the four key principles are addressed in the **Web Site Design Method (WSDM)**, except for iteration. User participation is incorporated into various stages, such as in planning, analysis and design with minimum rating; while usability is available with minimum and moderate rating in planning, analysis, implementation and design respectively, and real interaction is available with moderate ratings in the analysis and design. The strongest stage in WSDM is the design stage. This stage will help the designers to distinguish the future users or visitors of the website and gain more information about their characteristics.

After reviewing the methodologies for developing web sites, extra stages are collected from these methodologies (see Table 6.11). The main focuses of these extra stages are: usability, navigation, promotion, prototyping and identifying user types. These stages are very significant for developing web sites. Therefore, most of these stages will be taken into consideration by the researcher to be added to the new participative framework for developing websites.

Table 6.11 Extra stages from methodologies for developing Web sites (Prepared by Tomayess Issa)

Methodology (developing web sites)	Extra stages
Human factor methodology for designing websites	Usability goals development
Relationship Management Methodology (RMM)	Hardware selection; navigation design and user interface
The W3DT design methodology	Navigation design and building a hypermedia application
Information development methodology for the web	Promotion and prototyping "is available under the Implementation phase"
The Web Site Design Method (WSDM)	User modeling and conceptual design

Table 6.12 demonstrates the strongest stages from methodologies for developing web sites, and presents the rating availability for the four key principles in each stage. It was noticed that all the stages were covered in the methodologies for development of web sites as the main focus for these methodologies are:

- Defining the users' goals and understanding the environment very precisely in order to meet the users' needs and analyze the cost benefits;
- Defining the materials to identify user tasks and how information should be maintained;
- Defining the navigational path between the entities in the website, designing of screen and button layouts, generating a first trial product of the system, and defining user usability requirements and their characteristics;
- Releasing the first sketch of the website that will be checked by representative users in order to ensure that it complies with the user requirements;
- Making the website more interesting and attractive so that more users visit it, via content providers contributing up-to-date information to the site.

6.5   Marketing Methodologies

This section will examine the actual values added by Marketing Methodologies and the benefits they will bring to the e-commerce framework, especially in developing websites. In this section, the researcher will examine several methodologies from the marketing perspective such as e-Marketing Plan, and will review methodologies, which were created by companies, which are developing websites for marketing. At the end of each methodology section, the researcher will present a table showing: (1) how the four key principles are addressed in each stage within the methodology; (2) the strongest stage for each methodology for developing web sites; and (3) the extra stages of each methodology. These extra stages will help the researcher to develop a more comprehensive structure for the new participative methodology for developing marketing websites.

Table 6.12 Summary of Strongest Stages from Methodologies for Developing Web Sites (Prepared by Tomayess Issa)

Stage	Methodologies for developing web sites	Principles			
		User participation	Usability	Iteration	Real interaction
Planning	Human Factor Methodology for Designing Websites (HFMDW)	1	2	1	0
	Relationship Management Methodology (RMM)	0	0	1	0
	The Web Site Design Method (WSDM)	1	2	0	0
Analysis	Human Factor Methodology for Designing Websites (HFMDW)	0	3	1	2
	The Web Site Design Method (WSDM)	1	2	0	2
Design	Relationship Management Methodology (RMM)	0	0	2	0
	The W3DT design methodology	0	0	0	0
	The Web Site Design Method (WSDM)	1	2	0	2
Testing	Human Factor Methodology for Designing Websites (HFMDW)	1	3	2	0
Implementation	Information development methodology for the web	1	0	0	0
Evaluation	Human Factor Methodology for Designing Websites (HFMDW)	0	3	2	0
Maintenance	Human Factor Methodology for Designing Websites (HFMDW)	0	0	1	3
	Information development methodology for the web	0	0	0	1

6.5.1   E-Marketing Plan

The E-Marketing plan is a "guiding, dynamic document that links the firm's e-business strategy with technology-driven marketing strategies and lays out details for plan implementation through marketing management" (Strauss et al. 2003, p. 46). The main ideas behind an e-Marketing plan are: (1) to achieve an effective and efficient e-business objective; (2) to increase revenues and reduce costs; (3) to serve "as a roadmap to guide the direction of the firm, allocate resources, and make tough decisions at critical junctures" (Strauss et al. 2003).

Strauss et al. (2003) suggest that there are two common types of e-marketing plans: the 'napkin plan' and the 'venture capital plan'. The former approach is to just "jot ideas on a napkin over lunch or cocktails and then run off to find financing" (Strauss et al. 2003, p. 47). However, these plans work only sometimes. While the latter plan basically focuses on building a suitable business plan to increase the profit and reduce the cost. Therefore, the traditional marketing plan needs to be introduced to define and clarify key questions about topics such as capital, new customers, product and service, pricing and customer support required to retain the customers. Sound planning and "thoughtful implementation are needed for long-term success in business" (Strauss et al. 2003).

The E-Marketing plan is divided into seven steps:

- **Situation Analysis:** this step will help the Marketers to define and review the firm's environment and involves SWOT (strengths, weakness, opportunities, and threats) analyses. Strengths and weakness of the company's internal situation need to be identified, new opportunities need to be defined to improve the current situation of the company, while the threats "are areas of exposure" (Strauss et al. 2003, p. 50). Also under this step, a review and analysis of the existing marketing plan needs to be carried out to identify appropriate strategies, objectives, and performance metrics for e-business.
- **E-Marketing Strategic Planning:** this step involves "determining the fit between the organization's objectives, skills and resources and its changing market opportunities" (Strauss et al. 2003, p. 51). Additionally, the Marketers will create a sustainable e-marketing strategy for the e-business goals from "marketers design segmentation, targeting, differentiation, and positioning strategies" (Strauss et al. 2003). This includes demographics, geographic location, psychographics, and behavior of potential customers. This information will help the marketers to formulate the e-marketing objectives.
- **Objectives:** three main issues need to be defined in an e-marketing plan: task (what one is planning to achieve by building this e-business); measurable quantity (how much); and time frame (setting a time to accomplish the e-business job).
- **E-Marketing Strategies:** in this step, the marketers need to identify the 4Ps (product, pricing place and promotion) and the relationship management requirements to "achieve plan objectives regarding the offer" (Strauss et al. 2003, p. 53). Product: What is planned to be produced at the end (by building the e-business) in terms of service, information, selling products or advertising; Pricing: what it will cost for the e-business to be implemented; Place: the location of the

e-business work; Promotion: the techniques that will need to be adopted in order to promote the e-business work. The relationship management strategies need to identify how to "build relationships with a firm's partners, supply chain members, or customers" (Strauss et al. 2003, p. 57). Some companies use Customer Relationship Management (CRM) or Partner Relationship Management (PRM) approaches. PRM software is used to build and develop a complete database, which retains information about business partner capabilities and communication. While the purpose of the CRM software is "to retain customers and increase average order values and life time value" (Strauss et al. 2003, p. 57).

- **Implementation Plan:** the marketers select the 4Ps, relationship management strategies, and other tactics to achieve the e-marketing objectives and to develop the implementation plan. To achieve the implementation plan, the firm needs to check if the following aspects are available to accomplish the firm's objectives "staff, department structure, application service providers, and other outside firms" (Strauss et al. 2003, p. 57). Furthermore, special tactics will be used in the website to collect information about users who are dealing with it, such as forms, feedback e-mail, and online surveys. According to Strauss et al. (2003), additional tactics, which can be used to collect information, include: "1) Web site log analysis software helps firms review user behavior at the site and make changes to better meet the needs of users, 2) Business intelligence uses the Internet for secondary research, assisting firms in understanding competitors and other market forces".
- **Budget:** the key aspect of this stage is to identify the expected costs and returns from the investment. Returns are matched "against costs to develop a cost/benefit analysis, ROI calculation, or internal rate of return (IRR)" (Strauss et al. 2003) to determine if it is worthwhile to continue with the project. Furthermore, during the implementation stage, the marketers observe whether the results (cost and revenue) are on the correct track for achieving the predicted cost/benefit ratio.
- **Evaluation Plan:** is used to evaluate the success of the website. The tracking system should be available before activating the website. "E-marketers use tracking systems to measure results and evaluate the plan's success on a continuous basis" (Strauss et al. 2003, p. 60).

This e-marketing plan is a very important tactic for the marketers to gain more information about the current situation of the business before releasing the new version of e-business. However, this plan lacks a few stages which need to be available in order to achieve user exceptions and requirements, such as design, testing, iteration and maintenance.

The strongest stages in the E-Marketing Plan are E-Marketing Strategies (under the planning stage), the implementation stage and the evaluation stage. E-Marketing Strategies will allow the designer to identify the 4Ps: product, pricing, place and promotion, and the relationship management requirements to achieve plan objectives for the website. In the implementation stage, the marketers will utilize the 4Ps, the relationship management strategies, and other tactics to achieve the e-marketing objectives. The evaluation stage involves tracking systems to measure results and evaluate the plans for the website.

Table 6.13 indicates that usability and iteration have zero ratings for this methodology. User participation is available in the planning and implementation stages

Table 6.13 E-marketing plan (Prepared by Tomayess Issa)

Stages / Principles	Planning	Analysis	Design	Testing	Implemen- tation	Evaluation	Mainte- nance	Extra stages
User participation	1	0	0	0	1	0	0	E-marketing strategies objectives and budget
Usability	0	0	0	0	0	0	0	
Iteration	0	0	0	0	0	0	0	
Real interaction	0	0	0	0	0	3	0	
Strongest stage in E-marketing plan	☑	–	–	–	☑	☑	–	

with minimal rating, and real interaction is available in the evaluation stage with maximum rating. To formulate the e-marking objectives, the marketers will collect general information about the users such as demographics, geographic location, psychographics and behavior of potential users in the planning stage, while in the implementation stage, special tactics will be used to collect information about the users such as forms, feedback e-mail, and online surveys.

6.5.2   The Adventures Company Methodology

The Adventures Company released a process methodology to enhance the development of websites from a marketing perspective in 2004. This methodology has five stages, each of which should be completed before moving to the next stage.

- **Orientate:** this stage will help the designers to know why they are developing this website. In this stage, the designer will define the following concepts: the goals, product details, and competition. These concepts will also help to determine the cost and time for establishing this website.
- **Blue Print:** this stage will produce the first sketch for the website, where the "marketing, technology and creativity collide; banging heads and eventually coming upon the best way to mix all three aspects and create the optimum product" (Adventures 2004).
- **Model:** this stage will combine the technology possibilities and the creativity from the sketch to produce the working model.
- **Build:** during this stage, the designers will build up the new system and make sure that the proposed website is tested repeatedly until it meets users' requirements.
- **Maintain:** through this stage, the website will be maintained in order to "continue functioning at optimum levels" (Adventures 2004).

From the Adventures company point of view, this methodology will meet the users' requirements when building a website from the marketing perspective; however, not all the possible stages are available in this methodology. When compared with other system development processes, it lacks detailed design.

Table 6.14 The Advertures Company Methodology (Prepared by Tomayess Issa)

Stages / Principles	Planning	Analysis	Design	Testing	Implemen- tation	Evaluation	Mainte- nance	Extra stages
User participation	0	0	0	0	0	0	0	Blue print and model
Usability	0	0	0	0	0	0	0	
Iteration	0	0	0	2	0	0	0	
Real interaction	0	0	0	0	0	0	0	
Strongest stage in The Advertures Company Methodology	–	–	–	☑	–	–	–	

Table 6.14 shows that user participation; usability and real interaction have zero rating; while iteration is available in the testing stage with moderate rating to ensure that the website is tested repeatedly until it meets users' requirements. The strongest stage in this methodology is testing, which allows the designer to test the project repeatedly until it meets users' requests and desires.

6.5.3 The Market-Vantage (Internet Performance Marketing) Methodology

The Market-Vantage Company introduced a new methodology process for developing websites to enhance the strategy of the websites in order to "reduce cost, increase customer loyalty and market analysis" (Market-Vantage 2003). This methodology has four stages, each of which should be completed before moving to the next stage.

- **Internet marketing goals, target markets, and strategy:** this stage helps the designers to ask the following questions in the planning process: what are you selling? Who are the buyers? Who are your competitors? In addition, how can potential customers find the product? (Market-Vantage 2003). Answers to these questions will give the designer a full picture of the purpose behind building this website.
- **Define/Refine Internet Marketing Strategy:** this stage helps the designers in two aspects: learning about users [the purpose behind the visit and tracking their visit]; and how the business will be enhanced by using the Internet for introducing the new products.
- **Implementation:** this stage establishes the website so that the users can start using the new product and check if it meets their requirements.
- **Measurement:** is part of ongoing maintenance of the website and checking if the results of using the website are meeting its goals, using software to track current and new users. Continuing support and recommendations are available from the designer to the website manager.

We notice that this methodology includes iteration, so as to ensure that the website is meeting the user requirements and providing appropriate company outcomes. However, this methodology is missing a few stages such as detailed analysis and design. These stages are imperative in developing a website so that the website achieves the goals of e-business and as well as meeting users' requirements.

Table 6.15 identifies that user participation and usability have zero rating (except in the maintenance stage), while iteration is available in the implementation stage with moderate rating to ensure that the website meets users' requirements. Real interaction is available in the maintenance stage with moderate rating to check if the website meets users' requirements and needs after the changes have been made.

The strongest stages in the Market-Vantage (Internet Performance Marketing) Methodology are the planning, implementation and the maintenance stage, which is under the measurement stage. The planning stage will help the designers to identify the purpose behind building the website, namely, the products/service being sold, the firms' competitors and buyers, and how to find the product via the web. The implementation stage is important in Market-Vantage to allow users to use the new product and to check if it meets their requirements. User information is used in the maintenance stage to review on-going performance of the website.

6.5.4 EnSky's Unique Methodology

EnSky Company initiated a new methodology for developing websites from the marketing perspective. This methodology has into nine stages, each of which should be completed before moving to the next stage.

- **Evaluation Overview:** this methodology divides the evaluation aspect into two types: pre-and post-evaluation. The former is a phase to define the user needs and

Table 6.15 The Market-Vantage (Internet performance marketing) methodology (Prepared by Tomayess Issa)

Stages Principles	Planning	Analysis	Design	Testing	Implementation	Evaluation	Maintenance	Extra stages
User participation	1	0	0	0	0	0	2	Define/refine Internet marketing strategy; and measurement
Usability	0	0	0	0	0	0	0	
Iteration	0	0	0	0	2	0	0	
Real interaction	0	0	0	0	0	0	2	
Strongest stage in The Market-Vantage methodology	☑	–	–	–	☑	–	☑	

requirements for success and to determine the approach to be used in the latter stages, namely to define "the methods to track the results in post-evaluation" (EnSky 1997). The initial pre-evaluation stage establishes the goals of the project and identifies the existing branding, "marketing strategies, middle market demographics, competitors, and developing an understanding of the business and sales models" (EnSky 1997). According to EnSky's methodology, the post-evaluation process is very useful to measure the effectiveness of the site against the goals, which were set in the pre-evaluation.

- **Design:** during this stage, the designer will define the specifications and requirements and document the design of the look of the "end product that extends from the branding and marketing strategies already employed" (EnSky 1997).
- **Develop:** this stage will carry out the outcomes from the design phase to build the website by using various tools such as templates and graphical files, which were created in the design stage.
- **Testing:** during this stage, the prototype website will be tested to determine if it meets the requirements of the users. According to the EnSky methodology, once the "testing requirements have been met and approved by the client the project is ready for deployment" (EnSky 1997).
- **Deployment:** during this stage, the designer will transfer all the files of the website to the in-house web server. After this stage, the designer will follow the methodology by using the promotion and maintenance stages so as to begin "the process of both updating the content on the site to keep it relevant, and marketing the site to create awareness and drive traffic to it ensuring ultimate ROI" (EnSky 1997).
- **Promote:** this stage will help to promote the website to the public, by using various tools such as press releases, link building, banner ad campaigns, and paid search engine or directory listing campaigns. These processes will be repeated from time to time in order to make sure that the promoting phase is effective.
- **Maintain:** via this stage, the designer will make sure that the website is updated and maintained regularly and facilitates "the adoption of global technological advances" (EnSky 1997).
- **ROI:** this stage reviews the cost and investment of developing the website and compares it with likely returns.
- **Measurement:** is part of the ongoing maintenance of the website, and is integral in determining the ROI. According to EnSky, various types of tools are used for these measurements such as, "search engine ranking and website visitor statistics, tracking sales, new customers etc." (EnSky 1997).

This methodology contains most of the stages, which are needed for the designer to develop a website which meets the e-business objectives, and to evaluate the returns against the costs. However, two stages are missing – detailed analysis and iteration.

Table 6.16 indicates zero rating for the four key principles except for minimal user participation in the testing and maintenance stages and a minimal rating for real interaction in the maintenance stage. This means that the four key principles are

Table 6.16   EnSky's unique methodology (Prepared by Tomayess Issa)

Stages / Principles	Planning	Analysis	Design	Testing	Implemen-tation	Evaluation	Mainte-nance	Extra stages
User participation	0	0	0	1	0	0	1	Develop; ROI; measurement; and promotion
Usability	0	0	0	0	0	0	0	
Iteration	0	0	0	0	0	0	0	
Real interaction	0	0	0	0	0	0	1	
Strongest stage in EnSky's unique methodology	–	–	–	–	–	–	☑	

mainly ignored in this methodology. The strongest stage is maintenance (under the measurement stage). This stage is important to the designer and users simultaneously, as it will attract more users to visit the site. In addition, this stage includes changes and correction of errors in hardware and software to meet user requirements.

6.5.5   Review of Marketing Methodologies

The analysis above indicates that most stages in the marketing methodologies are similar to those in lifecycles, methodologies, and models, with extra stages focusing on the marketing perspective, such as measurement, promotion and cost/benefit analysis. These extra stages will help the firm to achieve "its desired results as measured by performance metrics according to the specifications of the e-business model and e-business strategy" (Strauss et al. 2003, p. 60).

6.5.6   Summary of Marketing Methodologies

This section will provide a summary behind the Marketing Methodologies
   E-Marketing Plan usability and iteration have zero rating while user participation is available in planning and implementation with minimal rating to collect general information about the users. Real interaction is available in the evaluation stage with maximum rating as the e-marketers use tracking systems to measure the results and ensure that the website meets users' requirements. The strongest stages in E-Marketing Plan are E-Marketing Strategies, implementation and evaluation. E-Marketing Strategies will allow the designer to identify the 4Ps: product, pricing, place and promotion, and the relationship management requirements to achieve plan objectives for the website. To achieve the implementation stage, the firm needs to check if all the objectives are available to accomplish the firm's needs. The evalu-

ation stage is for tracking the users' behaviors to establish whether the website meets their requirements.

In **the Advertures Company Methodology**, user participation, usability, and real interaction have zero rating, while iteration is available in the testing stage with moderate rating. Testing is the strongest stage in this methodology as this allows the designer to test the project frequently until it meets users' requests and desires.

The **Market-Vantage (Internet Performance Marketing) Methodology** is similar to the Advertures Company Methodology, as user participation and usability have zero rating (except for a moderate rating for participation in the maintenance stage). Iteration can be found in the implementation stage to ensure that the website meets users' requirements. Real interaction is available in the maintenance stage. The strongest stages are planning, implementation, and maintenance. The planning stage will allow the designers to gain more information about the rationale behind building the website; i.e. what is being sold; the firm's competitors and buyers; and how to find the product via the web. The implementation stage will allow the users to use the new product and check if it meets their needs. User satisfaction is tested during the maintenance stage.

The **EnSky's Unique Methodology** has zero ratings for the four key principles, except for a minimal rating for participation in the testing stage and real interaction in the maintenance stage. The strongest stage in EnSky's Unique Methodology is maintenance. This stage involves ongoing changes and correction of errors in hardware and software, in order to continue to meet user requirements.

After reviewing the marketing methodologies, extra stages were identified (see Table 6.17), focusing mainly on: promotion, prototyping, budget, ROI (return on investment) and measurement. These stages are important for developing websites from the marketing perspective. Therefore, the researcher will take into consideration these stages for the new participative framework for developing websites. The key techniques involved are:

- Identify the 4Ps for the E-Marketing plan: product, pricing, place and promotion;
- Identify the time frame to accomplish the job;
- Identify the expected returns from investment;
- Produce the first sketch for the website, evaluate it, then move on to produce the working model;
- Learn about the users by tracking their visit and the purpose behind the visit.

Table 6.17 Extra stages of marketing methodologies (Prepared by Tomayess Issa)

Methodology (marketing)	Extra stages
E-marketing plan	E-marketing strategies, objectives and budget
The Advertures Company Methodology	Blue print and model
The Market-Vantage (Internet performance marketing) methodology	Define/refine Internet marketing strategy and measurement
EnSky's unique methodology	Develop, ROI, measurement and promotion

Table 6.18 Summary of marketing methodologies

Stage	Marketing methodologies	Principles			
		User participation	Usability	Iteration	Real interaction
Planning	E-marketing plan	1	0	0	0
	The market-vantage(Internet performance marketing) methodology	1	0	0	0
Analysis	–	–	–	–	–
Design	–	–	–	–	–
Testing	The Advertures Company Methodology	0	0	2	0
Implementation	E-marketing plan	1	0	0	0
	The market-vantage(Internet performance marketing) methodology	0	0	2	0
Evaluation	E-marketing plan	–	–	–	3
Maintenance	The Market-Vantage methodology	2	0	0	2
	EnSky's unique methodology	1	0	0	1

Table 6.18 demonstrates the strongest stages for Marketing Methodologies and presents the rating for the four key principles in each stage. The main focuses of these methodologies are:

- Identify the product, pricing, place, promotion, and the relationship management requirements to achieve plan objectives for the website;
- Planning the purpose behind building the website; i.e. what are you selling; your competitors and buyers; and how to find the product via the web;
- Testing the website repeatedly until it meets users' requests and desires;
- Maintaining the website to attract more users (new as well as old) to visit it

6.6 Detailed Website Design and Implementation

The previous sections highlighted the need for a detailed approach to website design. This can lead to an effective website implementation, including organizational aspects. Two types of approaches will be discussed from the web-based hypermedia application perspectives in this section: The Object-Oriented Hypermedia Design Model and the Implementation Model.

6.6.1 The Object-Oriented Hypermedia Design Model (OOHDM)

Schwabe and Rossi (1995) describe an (Object-Oriented Hypermedia Design Model) OOHDM, a new model especially for designing a complex Web-based hypermedia application. The main aims of this approach are to: reduce complexity, make the website easy to navigate and maintain, thereby saving time and money, and make it more attractive to the users. This approach clearly separates the "navigational from conceptual design by defining different modeling primitives in each step" (Schwabe and Rossi 1995, p. 46). This approach is divided into four sequential stages, where each must be completed before the next stage can be started, although iteration can be used. Each stage "focuses on a particular design concern, and an object-oriented model is built" (Schwabe and Rossi 1995, p. 45). The stages are as follows:

- **Domain Analysis**: in this, stage the "conceptual model of the application domain is built using well-known object-oriented modeling principles" (Schwabe and Rossi 1995, p. 45).
- **Navigational Design**: in this stage the navigational structure for the hypermedia application will be defined in "terms of navigational contexts (focusing on the users and their tasks), which are induced from navigation classes such as nodes, links, indices, and guided tours" (Schwabe and Rossi 1995, p. 46).
- **Abstract Interface Design**: this stage provides the "perceptible objects" (i.e. picture, a city map … etc.) in "terms of interface classes" (i.e. text fields and buttons) (Schwabe and Rossi 1995, p. 46). Furthermore, this step will establish the communication between the interface and navigation in the hypermedia application.
- **Implementation**: In this stage, the hypermedia application will be implemented according to the user requirements and needs.

Table 6.19 illustrates that the design stage is very important for development of two key aspects of the website: navigational design and abstract interface design.

Table 6.19 The OOHDM methodology – extra stage (Prepared by Tomayess Issa)

The OOHDM methodology – extra stage
Design: two aspects will be defined in this stage: (1) navigational design; and (2) abstract interface design. The latter will define the navigational structure for the hypermedia application, while the former will establish the communication between the interface and navigation in the hypermedia application
Construction (Implementation): involves the technical implementation of the design

6.6.2 Implementation Methodology

Sampson et al. (2001) describe a methodology, which provides for the development of effective websites for counseling and career services. This methodology is very useful as it "can be used to consider opportunities for enhancing the design and use of the site" (Sampson et al. 2001) and it incorporates organizational aspects of implementation.

This model is divided into seven sequential stages, each of which must be completed before the next stage can be started. The stages are as follows:

- **Program evaluation:** this stage provides the foundation for the implementation process, helping to "ensures that the website is used for the right reasons with the right clients" (Sampson et al. 2001). The step has several sub steps: evaluate the current resources and services; establish a committee; prepare an implementation plan; and seek stakeholder support.
- **Web site development:** this stage helps the designer to make sure that the "web site developed has the potential to effectively meet client and organization needs" (Sampson et al. 2001). The stage has several steps: develop and evaluate website contents and features, and develop site documentation. In addition, this stage focuses on the development of website contents. Three questions need to be asked: "Whom does the website serve? What are the needs of users and what resources exist that would meet each of the identified needs?" (Sampson et al. 2001).
- **Web site integration:** this stage involves the users to make sure that the website outcomes will meet their requirements. It begins with the "staff reviewing current needs and current resources and services" (Sampson et al. 2001), and then determining how the website will be used in delivering services and how it will operate according to user requirements.
- **Staff training:** necessary training is given to the staff to incorporate the web site with existing service delivery.
- **Trial use:** this stage requires the users to try out the website to see if it meets their needs. Moreover, continuing training is available in this stage, and observation and interview methods are used in order to determine if the website training is effective.
- **Operation:** this stage allows the user to operate and use the website.
- **Evaluation:** evaluation and comments are collected from the users to ensure that the website services are running according to the user requirements. Therefore, the "results of the evaluation are used to indicate needed improvements in web site design and use" (Sampson et al. 2001).

Finally, the feedback loops are indicated by the arrows and the staff responds to feedback as the implementation process continues. It was noted that this model is most suited to the development of websites for counseling and career services. However, it also has a wider application. This method includes a stage, which is essential to the system development process, which is Training Staff (see Table 6.20).

Table 6.20 Implementation methodology – extra stage – prepared by Tomayess Issa

Implementation methodology – extra stage
Training Staff: from Implementation Model. This phase provides necessary training to the staff about the new system

6.7   Summary of Information Systems Development Methodologies, Methodologies for Developing Web Sites, and Marketing Methodologies

New challenges have been imposed since the growth of use of the Internet as a global means of delivering information, selling goods, and entertainment. These new challenges suggest the need to develop a new methodology for developing websites which meet users' requirements and needs in order to avoid potential client frustration, make the website enjoyable, effective and efficient, and most importantly, to improve performance.

In this section, the researcher will summarize the results from the earlier analysis of Information Systems Development Methodologies, Methodologies for Developing Web Sites, and Marketing Methodologies. The purpose behind the analysis is to:

- Identify the strongest stages of each methodology;
- Identify how well the four key principles are addressed in each methodology;
- Identify the extra stages from website and marketing methodologies.

Identifying the strongest stage for each methodology will help the researcher to define the framework for the new participative methodology for developing websites.

The researcher identified several stages from the development life cycle, which are: (1) planning, (2) analysis, (3) design, (4) testing, (5) implementation, (6) evaluation, and (7) maintenance. These stages are considered the basic and essential requirements for the system development process, as via those stages the designer will develop a system (interface or website) which meets the users' requirements.

Additionally, under the tables summarizing stages in the methodologies the researcher added four extra rows: "user participation," "usability," "iteration" and "real interaction." These key principles were either not fully considered in some methodologies, or were totally ignored. These principles are identified as being fundamental to the proposed system development process of a website for marketing purposes, producing an effective interface or website. Simultaneously, through these principles, the designer and user will develop the new system (interface or website) to meet the user requirements and needs in order to make the design system flexible and adjustable, and to limit user frustration when working with it. These principles are the main foundation for this research.

The first row is "**user participation**." It was noticed that user participation is a very practical approach in the development process. With it, the users will perform some activities and tasks and "these activities may pertain either to the management of the ISD project or to the analysis, design, and implementation of the system itself" (Hartwick and Barki 2001, p. 21).

Furthermore, according to Hartwick and Barki (2001), four dimensions of user participation can be identified: *RESPONSIBILITY; USER-IS RELATIONSHIP HANDS-ON ACTIVITY,* and the most important aspect, which is *COMMUNICATION ACTIVITY.* These dimensions can deliver the following information to the designer.

- **Responsibility**: "the performance of activities and assignment reflecting overall leadership or accountability for the project."
- **User-IS Relationship**: "the performance of development activities reflecting users' formal review, evaluation and approval of work done by the IS staff".
- **Hand-On Activity**: "the performance of specific physical design and implementation tasks."
- **Communication Activity**: "activities involving formal and informal exchange of facts, needs, opinions, visions, and concerns regarding the project among the users and between user and other project stakeholders" (Hartwick and Barki 2001, p. 22).

Therefore, the designer needs to work very closely with these dimensions in order to gain the basic information from the user about the system requirements and to identify the problems of the system. Furthermore, "user objectives, assumptions, strategies, actions, errors, problems, attitudes, etc., should surface so they can be explicitly considered in the system design and implementation processes" (Hartwick and Barki 2001, p. 22).

In addition, communication between the designers and users is an important aspect, which helps to identify the problems and to develop various solutions for the system by using different negotiation approaches and placing more emphasis on listening to users' needs and desires. For example, Joint Application Development (JAD) workshops are "facilitated by a session leader trained in group dynamic techniques, where users and developers work together to plan and design a new system" (Hartwick and Barki 2001, p. 22).

The second row is "**usability**." This term is very important in the system development process as usability involves "an assortment of support for needs such as ease of use, ease of learning, error protection, graceful error recovery, and efficiency of performance" (Carroll 2002, p. 193). Usability will be emphasized in this research as it is considered very important especially in a methodology for developing websites.

The third row is "**iteration**." This term is very important in the system development process, as it can occur in each stage to ensure that the web site is meeting the user requirements and company outcomes. This will enable the designers to build up the new website and make sure that the project will be tested repeatedly until it meets user requirements.

The fourth row is "**real interaction**." This term is very important in developing a website as it occurs in the maintenance and evaluation stages to ensure that user requirements are being met, by tracking use of the website by real users to achieve their specific objectives.

Finally, for the new participative framework for developing websites, a column will be added called "**participation rating**" which will help the researcher to identify the level of need for user participation in each stage. The participation rating will be from 0 to 3, indicating zero participation to maximum participation. The 1 and 2 ratings are minimum and moderate participation respectively

The researcher earlier reviewed the Mumford (1995) classification of user participation approaches in the system development process. In this research, the researcher will be using only the first two approaches: the Consultative Approach and the Representative Approach. Both of these approaches are very appropriate in all the stages in order to secure the agreement between users and designers at the beginning and to identify the key aspects, such as system objectives, problems, and the creating of various solutions to the system. The Consensus Approach will not be adopted in this research as it "does not always emerge easily and conflicts which result from different interests within a department may have to be resolved first" (Mumford 1995, pp. 18–19).

Extra stages were added from various methodologies for developing web sites, mainly focusing on: identifying user types, navigation, promotion, and prototyping. In addition, the researcher included more stages from marketing methodologies mainly focusing on: promotion, prototyping, budget, ROI (return on investment), and measurement.

The requirements of a new participative methodology for developing websites include:

- Participation at all stages (different participation rate);
- Provision of detailed contents acquisition and maintenance requirements;
- Provision for detailed design of presentation;
- Provision of usability evaluation (at various stages);
- Provision of regular maintenance.

Table 6.21 summaries the key aspects of the methodologies discussed in earlier sections of this chapter.

6.8   New Participative Methodology for Marketing Websites (NPMMW)

The New Participative Methodology for Marketing Websites (NPMMW) is developed from various existing models of system development and methodologies including lifecycle models, information systems development methodologies, methodologies for developing websites, marketing methodologies, and additional detailed techniques (see Figs. 6.1 and 6.2). .

Table 6.21 New participative *framework* for developing websites (Prepared by Tomayess Issa)

Stage	Partici-pation rating	Methodologies	Principles			
			User participation	Usability	Iteration	Real interaction
Planning	3	Soft system methodology	1	0	2	0
		Human Factor Methodology for Designing Websites (IIFMDW)	1	2	1	0
		Relationship Management Methodology (RMM)	0	0	1	0
		The Web Site Design Method (WSDM)	1	2	0	0
		E-marketing plan	1	0	0	0
		The Market-Vantage (Internet performance marketing) methodology	0	0	0	0
Analysis	2	Soft Systems Methodology (SSM)	3	0	2	0
		User Centered Development Methodology (UCDM)	1	0	0	0
		Ethics methodology	3	0	0	0
		Human Factor Methodology for Designing Websites (HFMDW)	0	3	1	2
		The Web Site Design Method (WSDM)	1	2	0	2
		Task analysis				
Design	3	Structured Systems Analysis and Design Methodology (SSADM)	1	0	1	0
		Soft Systems Methodology (SSM)	2	0	2	0
		User Centered Development Methodology (UCDM)	1	3	1	0
		Relationship Management Methodology (RMM)	0	0	2	0
		The W3DT design methodology	0	0	0	0
		The Web Site Design Method (WSDM)	1	2	0	2
		Navigation				
		Prototyping				

(continued)

Table 6.21 (continued)

Stage	Partici-pation rating	Methodologies	Principles			
			User participation	Usability	Iteration	Real interaction
Testing	3	User Centered Development Methodology (UCDM)	1	3	1	0
		Human Factor Methodology for Designing Websites (HFMDW)	1	3	2	0
		The Advertures Company Methodology	0	0	2	0
Implemen-tation	2	Information development methodology for the web	1	0	0	0
		E-marketing plan	1	0	0	0
		The Market-Vantage (Internet performance marketing) methodology	0	0	2	0
		Construction				
		Promotion				
		Staff training				
Evaluation	3	User Centered Development Methodology (UCDM)	2	3	1	0
		Human Factor Methodology for Designing Websites (HFMDW)	0	3	2	0
		E-marketing plan	0	0	0	3
		Measurement				
Mainte-nance	2	Human Factor Methodology for Designing Websites (HFMDW)	0	0	1	3
		The Market-Vantage methodology	2	0	0	2
		EnSky's unique methodology	1	0	0	1

Participation rate is from 0 to 3. Zero represents no participation while 3 indicates maximum participation. Ratings of 1 and 2 are minimum and moderate participation respectively. The ratings are based on the Consultative and Representative approaches according to Mumford (1995)

There are various comparisons with respect to the stages between methodologies for developing information systems, websites, or marketing strategies; however, integrating stages from information systems methodologies into a website with marketing methodologies is very valuable to improve websites that are more operative and effectual. User participation should be included in these methodologies to ensure that transaction processes, tracking, maintenance, and updating of the website meet the users' requirements.

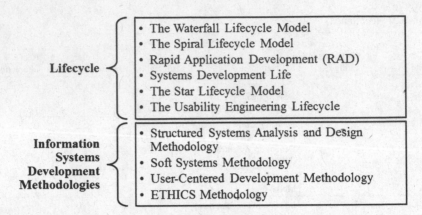

Fig. 6.1 Academic methodologies for development of websites

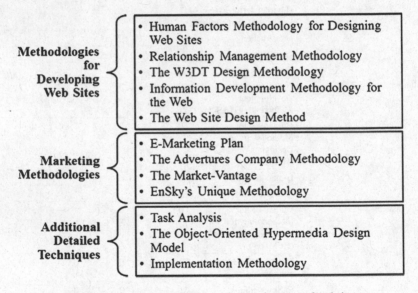

Fig. 6.2 Academic and commercial methodologies for development of websites

Each methodology was reviewed to determine two elements: (1) the stages needed for the system development process; and (2)the utilization of four key principles (user participation, usability, iteration, and real interaction (i.e. the monitoring of user interaction with aprototype site)). These principles were chosen to address the main deficits identified in existing website development methodologies, and to produce a new methodology, which will assist in the development of websites with high usability.

The major stages of the New Participative Methodology for Marketing Websites (NPMMW) are presented in Fig. 6.3. Table 6.22 shows the issues, tools and techniques for each stage and step, which need to be carried out by the designer in order

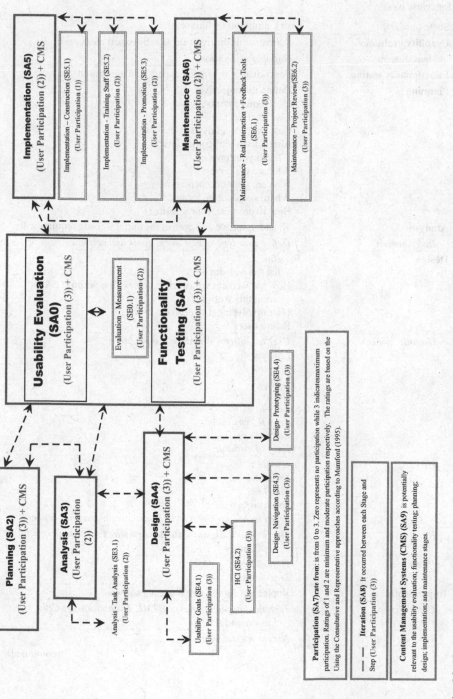

Fig. 6.3 New participative methodology for developing websites from the marketing perspective (Prepared by Tomayess Issa)

Table 6.22 Issues, tools and techniques for the new participative methodology (Prepared by Tomayess Issa)

Stage (*and step*)	Issues, tools and techniques
Usability evaluation	Formative usability evaluation by expert and user based
Measurement	*Ongoing evaluation*
Functionality testing	Functionality testing by expert- and user-based
Planning	Define the objectives User requirements User analysis Cost-benefits analysis Alternatives and constraints What is your product? Who are the buyers? Who are your competitors? Where should it be located? How to promote your website?
Analysis	To add, improve and correct the initial website requirements
Task Analysis	*Define user types, their work, goals and activities*
Design	To define: What the website is? How the website will work to achieve the purpose behind using this website? User involvement in decision making Future users
Usability goals	*User usability – Web design should be* *Efficient* *Effective* *Safe* *Utility* *Easy to learn* *Easy to remember* *Easy to use* *Easy to evaluate*
HCI goals	*Usable* *Practical* *Visible* *Job satisfaction* *Extra techniques, text style, fonts, layout, graphics and color*
Navigation	*Site, layout, link, navigational structure for the hypermedia application*
Prototyping	*High-fidelity* *Low-fidelity*
Implementation	Implementing the website using software
Construction	*Technical application (i.e. HTML, Dreamweaver; Cold Fusion and ASP)*
Training Staff	*Necessary training*

(continued)

Table 6.22 (continued)

Stage (*and step*)	Issues, tools and techniques
Promotion	*Press releases*
	Link building and banner-ad campaigns
	Paid search engine
	Directory listing campaigns to promote the website
	Traditional marketing (i.e. Newspaper; Radio and TV)
Maintenance	Update changes and the corrector of errors in the website
Real interaction +	*Log file*
feedback	*Forms, survey, discussion forum, contact form and telephone*
	number
Project review	*Checklists*

to achieve a user-friendly website to prevent user frustration when s/he deals with this interface. The major stages of the methodology may be described as follows:

Usability Evaluation (SA0): this stage is located at the center of the new methodology, as, before the process moves on to another stage, it is necessary to evaluate the results from the previous stage, which is known as "formative evaluation." *Usability Evaluation – Measurement (SE0.1):* this step is an ongoing evaluation of the website to ensure that it achieves its intended purposes.

Functionality Testing (SA1): this stage is also located at the center of the new methodology (with the usability evaluation) to test the results from the previous stage before moving to another stage. Expert-based and user-based evaluations will test the website to ensure that it functions effectively from the technical perspective.

Planning (SA2): this stage allows designers and users to address various project-scoping issues: (1) the requirements for developing a website; (2) the nature of the product and the buyers; (3) the firm's competitors; (4) the location of the site and how to promote the website. In addition, this stage involves developing a detailed schedule of activities required in order to carry out the development of the website in an efficient and effective manner.

Analysis (SA3): in this stage, users, analysts, and designers expand their findings in enough detail to indicate exactly what will and will not be built into the website design, and to add, improve, and correct the initial website requirements if they are not meeting the users' needs and wishes. Analysis – *Task Analysis (SE3.1):* this step will define the purpose of developing the website, the type of users, the type of work users will do with the website, users' goals, and their activities.

Design (SA4): the design stage will utilize the requirement specification from the previous stage to define: (1) what the website is; (2) how the website will work; (3) user involvement in decision-making; (4) future users; (5) usability requirements. *Design –Usability Goals (SE4.1):* this step will allow users (end-users and client-customer users), analysts, and designers (internal and external) to confirm that the website design is efficient, effective, safe, useful, easy to learn, easy to remember, easy to use and to evaluate, practical, and visible, and that it provides job satisfaction. *2 Design – HCI (SE4.2):* this step will allow users (end-users and client-customer users), analysts, and designers (internal and external) to identify that the website design is practical. There are many specific issues that need to be taken into consideration when designing website pages, such as text style, fonts, layout, graphics, and colour. *Design –Navigation (SE4.3):* this step will define the specific navigation paths through the website among the entities to establish the communication between the interface and navigation in the hypermedia application. *Design –Prototyping (SE4.4):* this step is essential in the website design process, to allow users and management to interact with a prototype of the new website, to suggest changes, and to gain some experience in using it. This step allows the management to reduce costs and increase quality through early testing.

Implementation (SA5): this stage involves the technical implementation of the website design. It allows users to use the new product and to check whether it meets their requirements. *Implementation –Construction (SE5.1):* this step involves the technical implementation of the website design. *Implementation –Training Staff (SE5.2):* this step will give the necessary training to the staff about the new website. *Implementation –Promotion (SE5.3):* this step will use various tools such as press releases, link building and banner-ad campaigns, paid search engines, directory listing campaigns, and traditional marketing methods (e.g. Newspapers, radio and TV) to promote the website.

Maintenance (SA6): this stage involves ongoing maintenance of the website, including updating changes and the correction of errors in the website. *Maintenance –Real Interaction and Feedback Tools (SE6.1):* During the maintenance stage, real interaction needs to be tracked by using the server log file. This information is very useful to the designers for improving and enhancing the structure and the functionality of the website in order to encourage more users to visit it. In addition, feedback tools should be available on the website to enable the users to contact the website owner for information or personal communication and to provide feedback about the website. For example, forms, surveys, discussion forum, contact form, telephone number, and a prize should be available on the website to encourage the users to provide feedback about the website. The first author recommends that, in order to prevent spam, the organization's e-mail address should not be made available on the website. *Maintenance –Project Review (SE6.2):* this step should be available to ensure that the website is working towards the project goals. This means that, after putting the website online, the designers need to check the website after 1 week to evaluate whether the website construction and structure are working according to the users' needs and requirements. One example of a tool that can be used for the project review is a checklist for the goals and objectives, usability and technical requirements.

User Participation (SA7): this aspect is a very important concept in the methodology, as the main purpose is to allow user participation in the website development process in order to gain more information about the problems and alternative solutions from the users and to familiarize them with the system before it is released. For each stage, there is a rating (from 0 to 3), which indicates the extent of user participation in the development process.

Iteration (SA8): this occurs between each stage and step in the New Participative Methodology for Marketing Websites, to check that the website does indeed meet users' (end users' and client-customer users') requirements and company objectives before moving to another stage.

Content Management Systems (CMS) (SA9): this aspect is relevant to the usability evaluation, functionality testing, planning, design, implementation, and maintenance stages in the New Participative Methodology for Marketing Websites. This tool will allow the users to manage the web contents by allowing them to add, edit, remove, and submit information by using various templates and workflows without needing any previous knowledge of the website editing tools.

6.9 Conclusion

This chapter has outlined the basic concepts behind Methodologies including: life-cycle models, IS development methodologies, methodologies with explicit human factors aspects, websites methodologies, marketing methodologies, and additional detailed techniques such as task analysis and detailed website design and implementation. The main focus has been on defining users' requirements and needs, planning, analysis, design, testing, implementation, evaluation and maintenance. These stages are very useful in any methodology, as, via them, the designer will make sure that the system is running according to the needs of users and the client organizations. In addition, four key principles (user participation, usability, iteration, real interaction) were identified as fundamental aspects to develop systems in an effective manner. The four key principles are the main foundation for this research.

Having reviewed the stages from a wide range of methodologies, the chapter concludes with an introduction to the New Participative Methodology for Developing Websites from the Marketing Perspective combining the most effective aspects of the methodologies.

References

Advertures (2004) Process methodology. http://www.advertures.cz/alt/index_en.php?cat=compan y&sub=methodology. Accessed 10 Aug 2014

Avison D, Fitzgerald G (1993) Information systems development: methodologies techniques and tools. Alfred Waller Ltd, New York

Avison DE, Wood-Harper AT (1990) Multiview: an exploration in information systems development. Alfred Waller Ltd, New York

Bichler M, Nusser S, Wien W (1996) Modular design of complex web-applications with W3DT. In: 5th international workshops on enabling technologies: infrastructure for collaborative enterprises (WET ICE'96), Standord, 1996. IEEE proceedings of WET ICE'96, pp 328–333

Boyer MA (1999) Step 1: satisfy the consumer. Supermark Bus 54(4):112

Carroll JM (2002) Human-computer interaction in the new millennium. Addison-Wesley, New York

Checkland P (1981) Systems thinking, systems practice. Wiley, Chichester

Checkland P (1988) Information systems and systems thinking: time to unite? Int J Inf Manag 8(4):239–248

Checkland P (2000) The emergent properties of SSM in use: a symposium by reflective practitioners. Syst Pract Action Res 13(6):799–823. http://link.springer.com/article/10.1023%2FA%3A1026431613200

Checkland P, Scholes J (1990) Soft systems methodology in action. Wiley, Chichester

Checkland P, Scholes J (2003) Soft systems methodology in action. Wiley, Chichester

Cropley DH, Cook SC (1999) Systems methodology for real-time information systems. University of South Australia. http://www.unisa.edu.au/sec/pubs/99papers/222-P192.PDF. Accessed 2 Mar 2015

Darlington K (2005) Effective website development. Pearson, Harlow

De Troyer O (1998) Designing well-structured websites: lessons to be learned from database schema methodology. In: Tok Wang Ling SR, Mong Li Lee (ed) Proceedings of the conceptual modeling – ER'98: 17th international conference on conceptual modeling, Singapore, 16–19 Nov 1998, Singapore, Springer-Verlag GmbH, pp 51–64

De Troyer O, Leune C (1998) WSDM: a user centered design method for web sites. In: Computer networks and ISDN systems, Proceedings of the 7th international World Wide Web conference, Elsevier, Vrijdag, 1998, pp 85–94

December J (1996) An information development methodology for the World Wide Web. Tech Commun 43(4):369

December J (2003) Developing information content for the world wide web. http://www.December.com/web/develop/overview.html. Accessed 22 June 2015

Dix A, Finlay J, Abowd G, Beale R (1998) Human-computer interaction, 2nd edn. Pearson Education Limited, Englewood Cliffs

Doll WJ, Torkzadeh G (1989) A discrepancy model of end-user computing involvement. Manage Sci 35(10):1151

EnSky (1997) EnSky's unique methodology. http://www.ensky.com/company/process/methodology.php. Accessed 10 Aug 2014

FFIEC IT Examination Handbook (2005) Systems development life cycle. Development and acquisition, vol 9, Nov 2005, USA

Flynn DJ (1992) Information systems requirements: determination and analysis. McGraw-Hill, New York/London

Hartwick J, Barki H (2001) Communications as a dimension of user participation. IEEE Trans Prof Commun 44(1):21–36

Hirschheim R, Iivari J, Klein H (1998) A comparison of five alternative approaches to information systems development. http://www.bauer.uh.edu/parks/fis/sad5.htm. Accessed 15 Feb 2012

Hix D, Hartson HR (1993) Developing user interfaces: ensuring usability through product & process. John Wiley & Sons, New York

Iivari J, Hirschheim R, Klein HK (2001) A dynamic framework for classifying information systems development methodologies and approaches. J Manage Inform Syst 17(3):179–218

Instone K (2004) Usability engineering for the web. W3C J. http://www.w3j.com/5/s3.instone.html. Accessed 22 June 2015

Isakowitz T, Stohr E, Balasubramanian P (1995) RMM: a methodology for structured hypermedia design. Commun ACM 38(8):34–44

Isakowitz T, Stohr E, Balasubramanian P (2000) RMM: a methodology for structured hypermedia design. http://www.dgp.toronto.edu/~fanis/courses/hypermedia/rmm.html. Accessed 2 July 2004

Issa T (2008) Development and evaluation of a methodology for developing websites. PhD thesis, Curtin University, Western Australia. http://espace.library.curtin.edu.au/R/MTS5B8S4X3B7SBAD5RHCGE

CEH2FLI5DB94FCFCEALV7UT55BFM-00465?func=results-jump-full&set_entry=000060&set_number=002569&base=GEN01-ERA02

Jayaratna N (1994) Understanding and evaluating methodologies -NIMSAD- a systemic framework. McGraw-Hill international, UK problem situation structured (2): at this stage, the analyst evaluates the problem situation from various approaches and different stakeholders; this means to examine and assess the situation from different worldviews. The stage has several steps: intervention analysis, social and cultural analysis, political analysis, rich picture and utilizes formal and informal methods

Laux L (1998) Designing web pages and applications for people with disabilities. In: Chris Forsythe EGJR (ed) Human factors and web development. Lawrence Erlbaum Associates, Mahwah, pp 87–95

Lazar J (2006) Web usability: a user-centered design approach. Pearson Education Inc, Boston

Market-Vantage (2003) Internet marketing methodology. http://www.market-vantage.com/about/methodology.htm. Accessed 10 Aug 2014

Mayhew DJ (1998) Human factors and the web. In: Chris Forsythe EGJR (ed) Human factors and web development. Lawrence Erlbaum Associates, Mahwah, pp 1–13

Mayhew DJ (1999) Business: strategic development of the usability engineering function. Interactions 6(5):27–34

McCracken DD, Wolfe RJ (2004) User-centered website development: a human-computer interaction approach. Pearson Education Inc., Upper Saddle River

Mumford E (1995) Effective systems design and requirements analysis. Macmillan, Great Britain

Mumford E (1996) Systems design: ethical tools for ethical change. Macmillan, Great Britain

Nielsen J (1992) The usability engineering life cycle. http://www-2.cs.cmu.edu/~jdh/courses/MOD/Neilesn-lifecycle.pdf. Accessed 1 Mar 2015

Olle TW, Hagelstein J, Macdonald IG, Rolland C, Sol HG, Assche FJMV, Verrijn-Stuart AA (1988) Information systems methodologies "A framework for understanding". Addison-Wesley Publishing Company, Boston

Preece J, Rogers Y, Sharp H (2002) Interaction design: beyond human-computer interaction. John Wiley & Sons, New York

Rosenhead J, Mingers J (2002) Rational analysis for a problematic world revisited. Wiley, Chichester

Sampson JP, Carr DL, Panke J, Arkin S, Minvielle M, Vernick SH (2001) An implementation model for Web site design and use in counseling and career services. The Florida State University, Tallahassee. http://www.career.fsu.edu/documents/implementation/Implementing%20Web%20Sites.ppt. Accessed 11 Sept 2011

Schwabe D, Rossi G (1995) The object-oriented hypermedia design model. Commun ACM 38(8):45–46

Strauss J, El-Ansary A, Frost R (2003) E-marketing, 3rd edn. Pearson Prentice Hall, Upper Saddle River

Vora P (1998) Human factors methodology for designing web sites. In: Chris Forsythe EGJR (ed) Human factors and web development. Lawrence Erlbaum Associates, Mahwah, pp 153–172

Chapter 7
New Participative Methodology for Sustainable Design (NPMSD)

Abstract Information and Communications Technology (ICT) use is increasing worldwide, since ICT has become a significant mechanism for researching, searching, communication, entertainment, shopping and information and more. However, the recycling of ICT products and the energy consumption of ICT is becoming a major problem for users and organizations nationally and internationally. Therefore, a solution should be applied to tackle and address it as a matter of urgency for the sake of the current and future generations. This chapter introduces and examines a New Participative Methodology for Sustainable Design. The sustainable design proposed in the New Participative Methodology for Sustainable Design was assessed via an online survey conducted in Australia. The survey outcomes confirmed the sustainable design step, and Australian users confirmed that through education and awareness, designers would learn more about sustainability and sustainable design.

7.1 Introduction

This chapter will introduce, discuss, and examine the new participative methodology for sustainable design. This methodology will assist designers to develop a new smart technology and portable devices with sustainability. Currently worldwide, the issues of recycling and energy consumption are causing a major dilemma by producing a carbon footprint, diseases and air pollution. Therefore, designers, academics, researchers, and individuals in general must understand their responsibility toward our planet. To tackle this problem, it is essential to raise designers' and HCI experts' awareness regarding their moral responsibility to create sustainable design for a sustainable future. Finally, our planet is suffering, and we need to tackle the issues of recycling, raw material supply and energy consumption, since there is no plan B for our planet. This chapter presents a new methodology for sustainable design in order to safeguard our planet. This chapter is organized as follows: introduction, New Participative Methodology for Sustainable Design, and conclusion.

© Springer-Verlag London 2015

T. Issa, P. Isaias, *Sustainable Design*, DOI 10.1007/978-1-4471-6753-2_7

7.2 New Participative Methodology for Sustainable Design

In Chap. 1, based on the literature review, the initial sustainable step identified six factors regarding sustainability, namely: design, safety, manufacture and energy, recycling, efficiency, and social needs. It is essential to consider these factors when developing a sustainable design for new smart technology and portable devices. Hence, designers and HCI experts should include these factors in their agenda to ensure that a good sustainable design will "eventually include criteria for the creation of a healthy environment and energy efficiency" (Stelzer 2006, p. 4).

Further studies (Dornfeld 2014; Funk et al. 2013; Mendler and Odell 2000; McDonough and Braungart 2002; McLennan 2004; Demirbas 2009; Wang et al. 2015a, b; Comm and Mathaisel 2015; Russell-Smith et al. 2015; Melles et al. 2015; Ramani 2010; Stelzer 2006) confirm that integrating sustainability in any business strategy including the design process, will enhance business reputation and preserve resources. Currently, sustainability and sustainable design are becoming the buzz words for users and organizations, as adopting and applying them in their strategies will be highly advantageous in terms of cost reduction, resources preservation, conformity to legislation, improvement of reputation, maintaining happier customers and stakeholders, attracting capital investment and capitalizing on new opportunities (Weybrecht 2010). Finally, Kendall and Kendall (2010) indicated that sustainability will assist businesses, education, stakeholders, individuals and society in general.

Today, the world population exceeds 7.2 billion, and by 2026, it will be more than 8 billion (Geoba.se; 2015). This increase will influence availability of housing, food, transportation, waste, economic, and social issues, employment the environment, and unsustainable development activities. Nowadays, there is an urgent call for sustainable development in all areas including new smart technology and portable devices. Hence, to tackle these problems, information technology and HCI experts should provide some solutions especially in design, manufacture, energy, waste management, and recycling by integrating and adopting sustainability and sustainable design in their design strategy especially for new smart technology and portable devices. This urgent call is essential to raise designers' and HCI experts' awareness regarding their moral responsibility toward sustainable development for a sustainable future.

A recent study by Kemp (2015) confirms that the number of active internet users is 3 billion, while mobile users is 3.6 billion; this means the yearly increase is around 21–5 % respectively. Table 7.1 shows the total number of active Internet users and mobile connections in various regions. These numbers are increasing daily, with a subsequent increase in the consumption of raw materials and the need for recycling.

Therefore, the issue of sustainability does not concern only the environment, but extends to social and economic issues. Using an appropriate methodology and smart technology for designing new smart technology and portable devices will enhance energy efficiency and reduce environmental impacts. Currently, increased usage of

Table 7.1 Digital usage by region (Kemp 2015)

Region	Total population [Million]	Active internet users [Million]	Mobile connections [Million]
Asia Pacific	4,021	1,407	3,722
Africa	1,135	298	900
Americas	979	633	1,068
Europe	837	584	1,104
Middle East	238	87	294

technology is becoming a pressing issue in the work since technology has a huge impact on the environment in terms of utilization of enormous amounts of raw materials, energy consumption, production of greenhouse gases and generation of electronic waste that harm both the planet and mankind, causing serious diseases and death (Gunn 2010; Philipson 2011; Shaw et al. 2015; Stewart and Kennedy 2009; Wiens 2013)

Individuals and organizations should understand that there is not another Earth to provide us with the essential resources and raw materials for our survival. Consequently, designers, users, and organizations should be mindful of the impact of their operations on the environment and take measures to become sustainable by integrating sustainability and sustainable design in their methodologies and strategies to reduce energy consumption and waste production, and keep in mind the importance of recycling.

Further, it is essential for designers, users, and organizations to reorient their methodologies and strategies towards sustainable design and sustainability considering the environment problems that the world is currently facing. Finally, it is important for users, organizations, HCI experts, designers to understand the impacts of their operations on the earth, particularly the technology use. Therefore, it is it is fundamental to take initiatives to address such problems by using innovative and creative sustainable solutions by educating users, organisations, HCI experts, designers, as well as top management about the importance of sustainable design and sustainability methodologies and strategies, which will increase technology performance and efficiency and reduce carbon emission as well.

This book attempts to address these issues by introducing a new sustainable model to tackle the new smart technology and portable devices design that can be applied now and in future. Therefore, the New Participative Methodology for Sustainable Design meets the needs of the present generation without compromising the needs of future generations. For example, research indicates that adopting this strategy offers various benefits in electric power reduction consumption of IT hardware and reduces CO2 emissions. Furthermore, it was indicated by Erek et al. (2009, p. 1) that 'Google, for instance, operates about 450,000 servers consuming nearly 800 million kWh a year'. Moreover, Google's data centers use around 260 million watts of power, which accounts for 0.01 % of global energy. This power is enough to consistently power 200,000 homes (StorageServers 2015). Therefore, power consumption by large organizations and users in general is increasing at an alarming rate. Hence, a new methodology should be implemented and applied to prevent or mitigate the undesirable outcomes related to manufacturing and energy consumption, by tackling the issues of design, recycling, safety, efficiency, and social impacts. These factors are taken into consideration in the new participative methodology for sustainable design.

The New Participative Methodology for Sustainable Design is driven by the New Participative Methodology for Marketing Websites (NPMMW) (Issa 2008). NPMMW has been developed from various existing models of system development and methodologies including lifecycle models, information systems development methodologies, methodologies for developing websites, marketing methodologies, and additional detailed techniques. NPMMW is divided into ten stages namely: usability evaluation; functionality testing, planning, analysis, design implementation, maintenance, user participation, iteration, and content management systems. NPMMW is a contingency methodology as it allows users and designers to select the techniques, which best meet, the requirements of the website, since each website from the marketing perspective has a different goal and objectives. To meet these objectives, the development of the website requires particular experience and skills.

The New Participative Methodology for Sustainable Design will use the same principle in its various stages and includes a new step in the design mainly for the purpose of sustainability. This sustainability step addresses the issues of design, manufacture and energy, recycling, safety efficiency, and social impact (see Fig. 7.1).

The design factor aims to facilitate upgrades and recycling, and the addition of new software; most importantly, it ensures compliance with environmental standards and rules.

Fig. 7.1 Sustainable step – factors. (Prepared by Tomayess Issa)

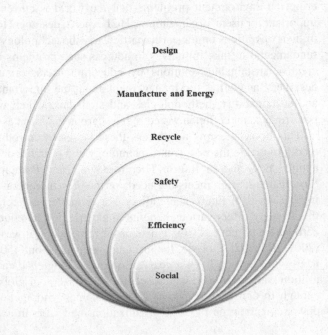

The safety factor aims to mitigate several negative outcomes of technology usage including carbon footprint, climate global warming, diseases, and air pollution. Therefore, the new smart technology design should consider these issues, especially in the recycling process.

Regarding the manufacture and energy factor, the new smart technology should tackle the energy issue by using less energy and raw materials, and produce less waste and toxins. Moreover, the new smart technology should use solar energy in the future.

In terms of the recycle factor, designers and HCI experts should use recycled, recyclable, and renewable materials to safeguard future supplies.

As for the efficiency factor, designers and HCI experts should develop new smart technology and devices with long life, less packaging and with portability efficiency.

Finally, regarding the social factor, it is desirable to shift the mode of consumption from personal ownership of products to provision of services, clean emissions, successful production cycles, and good ethical principles.

If the aforementioned factors are taken into consideration in new smart technology and portable devices design, resources for the next generation will be conserved, and our planet will be safeguarded from pollution, toxic emissions, and diseases. These factors come under the umbrella of the sustainability step in the design stage, which consists of usability goals, HCI, navigation, and prototyping.

The major stages and steps of the New Participative Methodology for Sustainable Design are presented in Fig. 7.2. Table 7.2 shows the issues, tools and techniques for each stage and step, which need to be carried out by the designer in developing a sustainable design. The major stages and stages of the New Participative Methodology for Sustainable Design may be described as follows:

Usability Evaluation (SA0) this stage is located at the center of the new methodology, as, before the process moves on to another stage, it is necessary to evaluate the results from the previous stage, which is known as "formative evaluation." Usability Evaluation – Measurement (SE0.1): this step is an ongoing evaluation of the new device to ensure that it will achieve its intended purpose(s).

Functionality Testing (SA1) this stage is also located at the center of the new methodology (with the usability evaluation) in order to test the results from the previous stage before moving to another stage. Expert-based and user-based evaluations will test the new device to ensure that it functions effectively from the technical perspective.

Planning (SA2) this stage allows designers and users to address various project-scoping issues: (1) the requirements for developing a new device, (2) the nature of the product and the buyers, (3) the firm's competitors. In addition, this stage involves developing a detailed schedule of activities required in order to carry out the development of the new devices in an efficient and effective manner.

Analysis (SA3) in this stage, users, analysts, and designers expand their findings in enough detail to indicate exactly what will and will not be built into the device

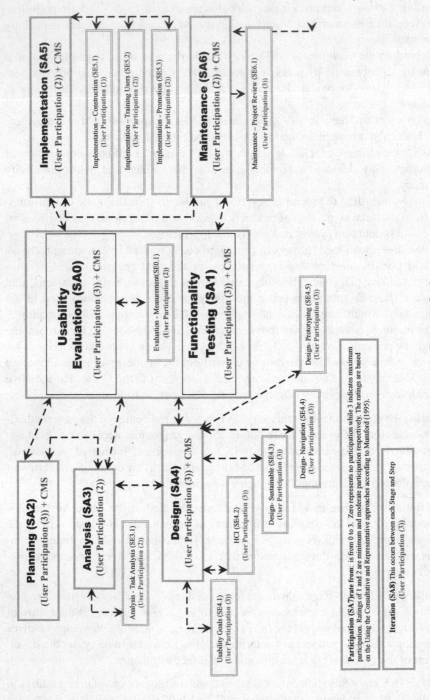

Fig. 7.2 New participative methodology for sustainable design (Prepared by Tomayess Issa)

Table 7.2 Stages, steps and issues, tools and techniques for the new participative methodology for sustainable design (Prepared by Tomayess Issa)

Stage (& *Step*)	Issues,tools and techniques
Usability evaluation	Formative usability evaluation by expert and user based
Measurement	*Ongoing evaluation*
Functionality testing	Functionality testing by expert- and user-based
Planning	Define the objectives User requirements User analysis Cost-benefits analysis Alternatives and constraints What is your product? Who are the buyers? Who are your competitors? Where should it be located? How to promote your smart technology or portable device?
Analysis	To add, improve and correct the initial smart technology or portable device requirements
Task analysis	*Define user types, their work, goals and activities*
Design	To define: What the smart technology or portable device is. How the smart technology or portable device will work to achieve the purpose behind using the new technology. User involvement in decision-making Future users
Usability goals	*User usability – smart technology or portable device design should be* *Efficient* *Effective* *Safe* *Utility* *Easy to learn* *Easy to remember* *Easy to use* *Easy to evaluate*
HCI goals	*Usable* *Practical* *Visible* *Job satisfaction* *Extra techniques, text style, fonts, layout, graphics and color*

(continued)

Table 7.2 (continued)

Stage (& *Step*)	Issues, tools and techniques

Sustainable	*Design*
	  Easy to upgrade
	  Easy to add new software
	  Easy to recycle
	  Sustain environmental standards and rules
	Safety
	  Reduce carbon footprint
	  Reduce global warming
	  Reduce diseases and even death of humans
	  Reduce air pollution
	  Reduce consumption and waste of resources
	Manufacture and energy
	  Use less energy
	  Use solar energy
	  Use less raw materials
	  Produce less waste and toxins
	  Recycle
	  Use recycled materials
	  Use recyclable materials
	  Use renewable materials
	Efficiency
	  Have long life
	  Have less packaging
	  Have portability efficiency
	Social
	  Shifting the mode of consumption from personal ownership of
	    products to provision of services
	  Have clean emissions
	  Have successful production cycles
	  Have good ethical principles

Navigation	*Site, layout, link, navigational structure for the hypermedia* *application*
Prototyping	*High-fidelity* *Low-fidelity*
Implementation	Implementing the smart technology or portable device
Construction	*Technical applications*
Training Users	*Necessary training*
Promotion	*Press releases* *Link building and banner-ad campaigns* *Paid search engine* *Directory listing campaigns to promote the smart technology or* *portable device* *Traditional marketing (i.e. Newspaper; Radio, and TV)* *Digital marketing (i.e. Internet and Social Networking)*
Maintenance	Update changes and the correct of errors
Project review	*Checklists*

design, and to add, improve, and correct the initial device requirements if they are not meeting the users' needs and wishes. Analysis – Task Analysis (SE3.1): this step will define the purpose of developing the device, the type of users, the type of work users will do with the device users' goals, and their activities.

Design (SA4) the design stage will utilize the requirement specifications from the previous stage to determine: (1) what the device is; (2) how the device will work; (3) user involvement in decision-making; (4) future users; (5) usability requirements. Design – Usability Goals (SE4.1): this step will allow users (end-users and client-customer users), analysts, and designers (internal and external) to confirm that the device design is efficient, effective, safe, useful, easy to learn, easy to remember, easy to use and to evaluate, practical, and visible, and that it provides job satisfaction. 2 Design – HCI (SE4.2): this step will allow users (end-users and client-customer users), analysts, and designers (internal and external) to identify that the device design is practical. There are many specific issues that need to be taken into consideration when designing a device, such as text style, fonts, layout, graphics, and colour. Design – Sustainable (SE4.3): this step will allow designers to consider the necessary factors for developing new smart technology and portable devices with sustainability in their agenda. Design – Navigation (SE4.4): this step will define the specific navigation paths through the device among the entities to establish the communication between the interface and navigation in the hypermedia application. Design – Prototyping (SE4.5): this step is essential in the device design process, to allow users and management to interact with a prototype of the new device, to suggest changes, and to gain some experience in using it. This step allows the management to reduce costs and increase quality through early testing.

Implementation (SA5) this stage involves the technical implementation of the device design. It allows users to use the new product and to check whether it meets their requirements. Implementation – Construction (SE5.1): this step involves the technical implementation of the new smart technology and portable device design. Implementation – Training Users (SE5.2): this step will give the necessary training to the users about the new smart technology. Implementation – Promotion (SE5.3): this step will use various tools such as press releases, link building and banner-ad campaigns, paid search engines, directory listing campaigns, and traditional marketing methods (e.g. Newspapers, radio and TV) and digital marketing methods (i.e. Internet and social networking) to promote the new smart technology.

Maintenance (SA6) this stage involves ongoing maintenance of the device. Maintenance – Project Review (SE6.1): this step ensures that the device is working towards the project goals. This means that, after the device is made 'alive', the designers need to check the device after 1 week to evaluate whether the device construction and structure are working according to the users' needs and requirements. One example of a tool that can be used for the project review is a checklist for the goals and objectives, usability and technical requirements.

User Participation (SA7) this aspect is a very important concept in the methodology, as the main purpose is to allow user participation in the device development process in order to gain more information about the problems and alternative solutions from the users and to familiarize them with the device before it is released. For each stage, there is a rating (from 0 to 3), which indicates the extent of user participation in the development process.

Iteration (SA8) this occurs between each stage, step in the New Participative Methodology for Sustainable Design, to check that the device does indeed meet users' requirements, and company objectives before moing to another stage.

To assess the Sustainable step including the factors, an online survey was conducted in Australia as a pilot study. The online survey was driven by the literature review and consists of two parts: background and sustainable design. The survey was distributed to the participants through the Qualtrics website (www.qualtrics. com). Qualtrics is an online survey tool that has a reliable reputation for developing and summarizing survey results; it allows users to complete online data collection and analysis (Boas and Hidalgo 2013). Table 7.3 shows the number and percentage of online survey participants in terms of gender, age, and qualifications. The survey response rate was 99.5 % and 51 % are female. The majority of respondents (15 %) were aged between 25 and 30 years, while the highest qualifications response rate is bachelor degree 27 %.

Table 7.4 shows the technology use by Australian participants. The online survey results confirmed that 42 % of Australian participants are spending up to 5 h per day on the computer for professional work and study; with 45 % on the Internet. Furthermore, 68 % spend less than an hour on email per day; while 69 % spend less than an hour daily on social networking.

Furthermore, the online survey confirmed that 93 % of the Australian users are using the Internet to access their email, while 75 % use it for banking online and 66 % for shopping online (see Table 7.5).

Furthermore, the online survey identified the devices used to access the Internet. Sixty-eight percent of respondents are using laptops, 61 % smartphones and 44 % use both PC and desktop (Table 7.6).

It came to our attention that Australian users were first introduced to the concepts of sustainability and green information technology via news media, school and Internet with 38 %, 33 %, and 32 % respectively (see Table 7.7).

The online survey examined the companies, which were associated with Australian users' devices. Apple and Google are the leaders at 33 % and 32 % respectively (see Table 7.8).

Furthermore, the survey asked Australian users whether they read the sustainability report of the manufacturer before buying a device. The survey shows that 53 % do not read the report; 23 % responded 'maybe'; and 11 % read the report. This outcome indicates that users should take more responsibility for their actions, and awareness of their responsibility to the planet needs to be raised via education and training (see Table 7.9).

In addition, Australian users change their device after 24–42 months with percentages ranging from 26 % to 20 % respectively (see Table 7.10).

Table 7.3 Online survey statistics Australia (Prepared by the authors)

Number and percentage of online survey	
Questionnaires distributed	209
Questionnaires returned	208
Response rate	99.5 %
Gender	
Male respondents	102 (49 %)
Female respondents	106 (51 %)
Age	
17 years and under	0 (0 %)
18–20	16 (8 %)
21–24	19 (9 %)
25–30	31 (15 %)
31–35	29 (14 %)
36–40	26 (12 %)
41–45	24 (11 %)
46–50	14 (7 %)
51–55	19 (9 %)
56–60	19 (9 %)
61–65	12 (6 %)
Over 65	0 (0 %)
Qualifications	
Primary education	4 (2 %)
Higher secondary/pre-university	37 (18 %)
Professional certificate	26 (13 %)
Diploma	30 (15 %)
Advanced/higher/graduate diploma	12 (6 %)
Bachelor's degree	56 (27 %)
Post graduate diploma	12 (6 %)
Master's degree	13 (6 %)
Ph D	6 (3 %)
Others	8 (4 %)

Table 7.4 Technology use by Australian users (Prepared by the authors)

Answer	Hours spend on the *computer* per day	Hours spend on the *internet* per day	Hours spend on the *email* per day	Hours spend on the *social networking* per day
	Response %	Response %	Response %	Response %
Less than an hour	22 (11 %)	40 (19 %)	142 (68 %)	144 (69 %)
Up to 5 h	87 (42 %)	93 (45 %)	61 (29.1 %)	55 (26 %)
5–10 h	69 (33 %)	47 (23 %)	3 (1.4 %)	3 (1 %)
10–20 h	26 (12 %)	25 (12 %)	3 (1.4 %)	5 (2 %)
Over 20 h	5 (2 %)	3 (1 %)	0 (0 %)	1 (2 %)

Table 7.5 Internet usage by Australian users (Prepared by the authors)

Answer	Response	%
Email	194	93
Playing games	82	39
Studying	80	38
Working	96	46
Shopping online	137	66
Chatting	79	38
Researching hobbies	101	48
Banking online	157	75
Buying goods or services	132	63
Buying stocks or investing online	24	11
Researching travel information or making reservations	105	50
Others – please specify	14	7

Table 7.6 Devices usage by Australian users. (Prepared by the authors)

Answer	Response	%
PC	91	44
Desktop	92	44
Laptop	143	68
Netbook	15	7
PDAs	4	2
Workstation	9	4
Tablet	43	21
Smartphone	128	61
Others – please specify	9	4

Table 7.7 First introduced to the concepts of sustainability and green information technology by Australian users. (Prepared by the authors)

Answer	Response	%
School	69	33
Higher education	37	18
Internet	67	32
Books	32	15
Magazine	31	15
News media	79	38
Conferences	9	4
Others – please specify	19	9

Table 7.8 Australian users devices (Prepared by the authors)

Answer	Response	%
Apple	67	33
Google	66	32
Dell	22	11
IBM	16	8
Others – please specify	35	17

Table 7.9 Australian users reading the sustainability report of the company before buying a device (Prepared by the authors)

Answer	Response	%
Yes	23	11
No	110	53
Maybe	48	23
Not at all	28	13

Table 7.10 Australian users changing their device (Prepared by the authors)

Answer	Response	%
Every 6 months	3	1
Every 12 months	16	8
Every 18 months	19	9
Every 24 months	54	26
Every 30 months	15	7
Every 36 months	33	16
Every 42 months	41	20
Other – please specify	27	13

Table 7.11 Australian users "Why do you change your device" (Prepared by the authors)

Answer	Response	%
Size	38	18
Speed	100	48
Functionality	116	56
Keeping with technology	110	53
Others – please specify	28	13

When asked by the survey "Why do you change your device?" the majority of participants (56 %) indicated that they changed because of the functionalities offered by the new device; 53 % want to keep up with technology, and 48 % want more speed (see Table 7.11).

Additionally, the online survey sought to determine the Australian users' attitudes to their moral responsibilities toward the planet by asking whether changing their devices frequently will cause damage to our planet. Table 7.12 indicated that

Table 7.12 Australian users "changing device frequently will cause damage to our planet" (Prepared by the authors)

Answer	Response	%
Yes	79	38
No	34	16
Maybe	93	44
Not at all	3	1

Table 7.13 Australian users: "Can we change the mindset of designers and users regarding sustainability" (Prepared by the authors)

Answer	Response	%
Training	96	46
Education	148	71
Awareness	140	67
Workshop	45	22
Internet	101	49
T.V.	91	44
Social networking	83	40
Others – please specify	10	5

44 % responded "Maybe" while 38 % showed their awareness that changing devices frequently would cause damage to our planet.

The online survey examined Australian users' recommendations of ways to change the mindset of designers and users toward sustainability. The survey concluded that via education and awareness (71 % and 67 % respectively) designers and users could change their mindset and attitude (see Table 7.13).

A total of *209 valid cases* were processed for the subsequent Factor Analysis. The analysis was conducted separately for two groups with 23 for sustainable design and 37 questions for the advantages and disadvantages of sustainability respectively. The first part for the group sustainable design consists of six groups/aspects based on users' level of awareness of sustainable design from Stelzer (2006, p. 4). Those aspects are design [4 questions]; safety [5 questions]; manufacture and energy [4 questions]; recycling [3 questions]; efficiency [3 questions]; and social factors [4 questions].

To further examine the online survey results, the researchers adopted principal axis factoring for factor extraction, and oblique rotation (rather than orthogonal rotation) was applied using the Promax method (Costello and Osborne 2005; Hair et al. 2009). To measure the sampling adequacy, researchers carried out specific testing using Cronbach's Alpha, Kaiser-Meyer-Olkin and Bartlett's test. Table 7.2 shows the statistical results for the Alpha, KMO and Bartlett's test. The Cronbach's Alpha for all 23 variables from sustainable design is .966, indicating an excellent internal consistency of the items in the scale (Gliem and Gliem 2003). A Kaiser-Meyer-Olkin (KMO) measure of sampling adequacy of .950 indicates an marvelous excellent sample size was obtained for the analysis (Hill 2012); hence, the current

KMO results are appropriate and acceptable for this study. The Bartlett's test of sphericity is highly significant, $\chi^2 = 4417.474$ $df = 253$, $p < .000$, indicating that the items of the scale are sufficiently correlated for factors to be found (Tobias and Carlson 1969; WIlliams et al. 2010). Therefore, results shown in Table 7.14 indicate the validity of the sustainable step.

Furthermore, the researchers used principle components analysis to estimate the factor-loading matrix for the factor analysis model as well the standard correlation matrix. The Eigen values were assessed to determine the number of factors accounting for the correlations amongst the variables. For the sustainable design section, Table 7.15 demonstrated the total variance with a total of 71.850 % of the variation. The Table 7.15 shows the variance is divided among the 23 component and indicated that three components to be extracted for these variables, the cumulative satisfy the criterion of explaining 60 % or more of the total variance as a three components solution would explain the 71.850 % of the total variance. The amount of variances explained by each of these components is presented in Table 7.15 (after rotation attempted).

KMO, Bartlett's test, and Alpha were used as measures for the sustainable design section (23 questions). An examination of the Kaiser = Meyer Olkin measure of sampling adequacy suggested that the sample was marvelous excellent (.951) and the Bartlett's test of Sphericity is $\chi^2 = 4175.429$, $df = 253 = p < .000$ which satisfies the sustainable design step; the Cronbach's Alpha sample was excellent at .969.

The communalities table represents the proportion of the variance in the original variables that is accounted for by the factor solution. The communality value for each variable is higher than 0.50.

In order to measure the regression coefficients (i.e. slopes), the researchers carried out the factor loadings. The factor loadings of most of the items were ade-

Table 7.14 Sustainable design – statistics (Prepared by the authors)

Sub group	Cronbach's alpha	KMO sampling adequacy	Bartlett's test of sphericity
Design	.849 [Good]	.726 (Middling)	$\chi^2 = 424.239$; $df = 6$ $p < .000$
Safety	.905 [Excellent]	.860 (Meritorious)	$\chi^2 = 701.911$; $df = 10$ $p < .000$
Manufacture and energy	.845 [Good]	.735 (Middling)	$\chi^2 = 374.469$; $df = 6$ $p < .000$
Recycle	.943 [Excellent]	.722 (Middling)	$\chi^2 = 568.083$; $df = 3$ $p < .000$
Efficiency	.844 [Good]	.718 (Middling)	$\chi^2 = 261.795$; $df = 3$ $p < .000$
Social	.871 [Good]	.815 (Meritorious)	$\chi^2 = 442.816$; $df = 6$ $p < .000$

Table 7.15 Total variance for sustainable design section (Prepared by the authors)

	Initial eigenvalues			Rotation sums of squared loadings		
Component	Total	% of Variance	Cumulative %	Total	% of Variance	Cumulative %
1	13.785	59.936	59.936	5.787	25.162	25.162
2	1.548	6.730	66.666	5.782	25.139	50.301
3	1.192	5.184	71.850	4.956	21.549	71.850
4	1.026	4.462	76.312			
5	.625	2.717	79.030			
6	.586	2.546	81.576			
7	.501	2.177	83.753			
8	.449	1.952	85.704			
9	.421	1.829	87.534			
10	.386	1.678	89.212			
11	.332	1.445	90.657			
12	.294	1.279	91.936			
13	.245	1.066	93.001			
14	.237	1.028	94.030			
15	.219	.954	94.984			
16	.205	.893	95.877			
17	.190	.826	96.703			
18	.160	.695	97.397			
19	.151	.657	98.054			
20	.130	.567	98.621			
21	.122	.532	99.153			
22	.110	.478	99.631			
23	.085	.369	100.000			

Extraction method: principal component analysis

quately high and the one with the cleanest fact structured to be considered as important (Costello and Osborne 2005), and to exclude several items under each factor where the factor loading is below 0.5 based on the rule of thumb of Stevens (1992) for a sample size above 100. In addition, Rose et al. (2011) suggested that the acceptable factor loading based on sample size between 200 and 249 is 0.40 (see Table 7.16). Table 7.16 shows the group pattern matrix for the sustainable design section.

The pattern Matrix revealed three factors, namely: (1) efficient resources, (2) reduce waste and resource, and (3) feasible design.

The online survey outcomes indicated that Australia is encouraging sustainable design for the current technology, new smart technology, and portable devices, by asking designers to integrate and adopt sustainability and sustainable design

Table 7.16 Rotated component matrix – sustainable design section (Prepared by the authors)

Rotated Component Matrix[a]

	Component		
	1	2	3
Use solar energy	.751		
Have successful production cycles	.729		.438
Use renewable materials	.660	.345	.416
Use recycled materials	.649	.366	.445
Have portability efficiency	.648		.485
Have good ethical principles	.617	.311	.442
Shifting the mode of consumption from personal ownership of products to provision of services	.612		
Use recyclable materials	.593	.364	.507
Have clean emissions	.556	.444	.483
Use less raw materials	.545	.520	
Have long life	.541	.340	.487
Reduce climate global warming		.830	
Reduce air pollution	.382	.809	
Reduce carbon footprint		.805	.381
Reduce consumption and waste of resources		.773	.331
Sustain environmental standards and rules		.691	.526
Reduce diseases and even death of humans	.535	.595	
Produce less waste and toxins	.445	.570	.496
Are easy to add new software			.839
Are easy to upgrade			.688
Have less packaging	.461	.387	.645
Are easy to recycle		.575	.627
Use less energy	.414	.494	.568

Extraction method: principal component analysis.
Rotation method: varimax with kaiser normalization.

[a]Rotation converged in nine iterations.

concepts in their design process. Australia wants designers to preserve raw resources and materials for future generations. Australian users confirmed that sustainable design is the way to go in the future, and we need to make users more aware of the consequences for future generations regarding sustainability by raising awareness through education and training, since the biggest problem is that most people seem to want the latest products on the market.

Finally, the survey outcomes confirmed the sustainable design step in the New Participative Methodology for Sustainable Design; this study assisted the authors to confirm their views regarding sustainable design.

7.3 Conclusion

This chapter is concerned with the development of New Participative Methodology for Sustainable Design, and identifying the sustainable design step, which comprises design, manufacture and energy, recycling, safety efficiency and social factors. This methodology was developed to raise designers and users' awareness of sustainability and green information technology in terms of technology and portable devices design. Using this methodology in designing devices and new smart technology will reduce the harm done to our planet as a result of poor recycling and the consumption of energy and raw materials. Finally, in order to raise awareness among users, we academics have a responsibility to increase our students' awareness, and make them part of the solution not the problem, encouraging them to become good stewards serving their countries and communities. In the future, additional research will be carried out to assess the sustainable design step using larger, more diverse countries with developed and developing economies to ensure compliance with environmental standards and rules for sustainable systems.

References

Boas TC, Hidalgo FD (2013) Fielding complex online surveys using rapache and qualtrics. Polit Methodol 20(2):21–26

Comm CL Mathaisel DF (2015) Designing an engineering system for sustainability. In applied mechanics and materials. Trans Tech Publ, pp 474–478

Costello A, Osborne J (2005) Best practices in exploratory factor analysis: four recommendations for getting the most from your analysis. Pract Assess, Res Eval 10(7):1–9

Demirbas A (2009) Political, economic and environmental impacts of biofuels: a review. Appl Energy 86:S108–S117

Dornfeld DA (2014) Moving towards green and sustainable manufacturing. Int J Precis Eng Manuf-Green Technol 1(1):63–66

Erek K, Schmidt N-H, Zarnekow R, Kolbe LM (2009) Sustainability in information systems: assortment of current practices in IS organizations. In: Americas Conference on Information Systems (AMCIS), San Francisco, pp 1–9

Funk B, Niemeyer P, Gómez JM (2013) Information technology in environmental engineering: selected contributions to the 6th international conference on information technologies in environmental engineering (ITEE2013). Springer Science & Business Media, Springer Heidelberg, New York/Dordrecht/London

Geoba.se (2015) The world population. http://www.geoba.se/population.php?pc=world&page=3&type=028&st=country&asde=&year=2026. Accessed 6 Jun 2015

Gliem JA, Gliem RR (2003) Calculating, interpreting, and reporting Cronbach's alpha reliability coefficient for Likert-type scales. In: Proceeding of the midwest research-to-practice conference in adult, continuing, and community education, Ohio, Ohio State University, pp 82–88

Gunn C (2010) Sustainability factors for e-learning initatives. ALT-J Res Learn Technol 18(2):89–103

Hair J, Black W, Babin B, Anderson R (2009) Multivariate data analysis. Prentice Hall, Upper Saddle River

Hill BD (2012) The sequential Kaiser-Meyer-Olkin procedure as an alternative for determining the number of factors in common-factor analysis: a Monte Carlo simulation. Umi Dissertation Publishing, Proquest

Issa T (2008) Development and evaluation of a methodology for developing websites – Ph D thesis, Curtin University, Western Australia. http://espace.library.curtin.edu.au/R/MTS5B8S4X3 B7SBAD5RHCGECEH2FLI5DB94FCFCEALV7UT55BFM-00465?func=results-jump-full&set_entry=000060&set_number=002569&base=GEN01-ERA02

Kemp S (2015) Digital, social and mobile worldwide in 2015. http://wearesocial.net/tag/sdmw/. Accessed 22 Jan 2015

Kendall K, Kendall J (2010) Forms of government and systemic sustainability: a positive design approach to the design of information systems. Adv Appreciative Inq 3:137–155

McDonough W, Braungart M (2002) Design for the triple top line: new tools for sustainable commerce. Corp Environ Strateg 9(3):251–258

McLennan JF (2004) The philosophy of sustainable design: the future of architecture. Ecotone Publishing, Kansas City

Melles G, Kuys B, Kapoor A, Rajanayagam J, Thomas J, Mahalingam A (2015) Designing technology, services and systems for social impact in the developing world: strong sustainability required. In: Chakrabarti AE (ed) ICoRD'15–research into design across boundaries, vol 2. Springer, pp 89–97

Mendler S, Odell W (2000) The HOK guidebook to sustainable design. John Wiley & Sons, New York

Philipson G (2011) ICT and sustainability. http://www.aiia.com.au/resource/collection/F0FF33B8-BF15-44B3-971E-20E03EC57287/Graeme_Philipson_AIIA_2011-09-29.pdf. Accessed 24 Jul 2012

Ramani K (2010) Sustainable design. J Mech Des 132:1–2

Rose S, Hair N, Clark M (2011) Online customer experience: a review of the business-to-consumer online purchase context. Int J Manag Rev 13:24–39

Russell-Smith SV, Lepech MD, Fruchter R, Littman A (2015) Impact of progressive sustainable target value assessment on building design decisions. Build Environ 85:52–60

Shaw G, Walters R, Kumar A, Sprigg A (2015) Sustainability in infrastructure asset management. In: Proceedings of the 7th world congress on engineering asset management (WCEAM 2012), Springer, pp 525–534

Stelzer K (2006) Sustainability = good design. Les Ateliters De Lethique 2:1–15

Stevens J (1992) Applied multivariate statistics for the social sciences. Erlbaum, Hillsdale

Stewart E, Kennedy J (2009) The sustainability potential of cloud computing: smarter design. http://www.environmentalleader.com/2009/07/20/the-sustainability-potential-of-cloud-computing-smarter-design/. Accessed 1 Dec 2009

Storage Servers (2015) Facts and stats of world's largest data centers. https://storageservers.word-press.com/2013/07/17/facts-and-stats-of-worlds-largest-data-centers/. Accessed 1 Jun 2015

Tobias S, Carlson JE (1969) Brief report: Bartlett's test of sphericity and chance findings in factor analysis. Multivar Behav Res 4(3):375–377

Wang L, Kwok JS, Tsang DC, Poon C-S (2015a) Mixture design and treatment methods for recycling contaminated sediment. J Hazard Mater 283:623–632

Wang L, Tsang DC, Poon C-S (2015b) Green remediation and recycling of contaminated sediment by waste-incorporated stabilization/solidification. Chemosphere 122:257–264

Weybrecht G (2010) The sustainable MBA – the manager's guide to green business. John Wiley & Sons, Chichester

Wiens JA (2013) Is landscape sustainability a useful concept in a changing world? Landsc Ecol 28:1047–1052

WIlliams B, Onsman A, Brown T (2010) Exploratory factor analysis: a five-step guide for novices. J Emerg Prim Health Care (JEPHC) 8(3):1–13

Chapter 8
Future ICTs: Present Trends for Future Developments

Abstract This chapter addresses Future ICTs, covering present trends and future developments. It is divided into two main sections: Social networks trends and Web 3.0 trends. Social Networks trends will detail aspects like the anonymity and privacy debate, Business, Education and other sectors. Web 3.0 trends will cover aspects like the semantic heterogeneity challenge, Business, Education and other sectors.

8.1 Social Networks Trends

This section addresses key trends regarding Social Networks (SNs), organized by broad categories, as depicted in Fig. 8.1 below.

In Table 8.1, it can be summarized some of the key trends detailed in the following sections.

8.1.1 The Anonymity and Privacy Debate

Some online social networks impose a real-name policy which prevents their users from using alias. This policy is justified by the social networks as a strategy to improve content and service, to facilitate users' search for contacts and to enable accountability. Despite the benefits that social networks often numerate to explain the adoption of this policy, there is a growing controversy associated with the use of the user's real identity. By requesting their users to register with their real identity, these platforms have access to information that jeopardizes privacy and online freedom (Peddinti et al. 2014). Users who are concerned with their privacy have found means to circumvent this policy. Also, some social networks, such as Twitter, do not condition users' registration to the use of their real identity (Peddinti et al. 2014).

The growth of health related social networks has raised issues of privacy for their users. While it has become known that the participation of patients in online platforms for health issues can represent an assortment of benefits, it can also pose a challenge in terms of the protection of the users' privacy. The authors developed a model that depicts the patients' information sharing behaviour based on three

T. Issa, P. Isaias, *Sustainable Design*, DOI 10.1007/978-1-4471-6753-2_8

Fig. 8.1 Social networks
broad categories of key
trends (Prepared by Pedro
Isaias)

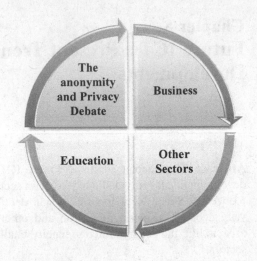

Table 8.1 Networks key trends (Prepared by Pedro Isaias)

Anonimity and privacy debate trends	Business in SNs trends
Alias vs. real-name policy	SN sites in business must follow specific guidelines
Volume of information vs. users privacy	Online SN presence fosters business relationships (both on-line as well as off-line)
Higher number of SN users – anonymity vs. de- anonymisation	SNs constitute great communication challenges
Transparency and the right to be forgotten	Key users roles in SNs is crucial
	SNs empower clients
Education in SNs trends	**Other sectors in SNs trends**
Distraction vs. positive role of SNs in Education	Health sector focus
Gender role of SNs in education	Social graph analyses
Mobile SNs usage	Sampling methods for SNs
	Citizen participation

factors: individual characteristics, type of information and breadth of the audience.
Patients seem to prefer moderate platforms that offer protection to their private
information, but facilitates the exchange of clinical data (Frost et al. 2014). This
model provides insight into important patterns of information sharing, which can
help to improve the design of online communities.

The rising number of social network users causes great volumes of varied information to be posted online. This volume of information is in its turn responsible for the
growing availability of datasets via the internet. Although users attempt to anonymise
their information, it is becoming increasingly uncertain if their data is protected from
de-anonymisation. In light of this predicament, transparency is rising as a new framework for information management. In addition to transparency, the right to be

forgotten is vital in information management in the sense that it would allow users to delete previous data, when introducing new information (Kataoka et al. 2014).

Transparency has the potential to endow social network users with the sense of increased control. This perception of control can have a positive impact on the effectiveness of online advertising in social networks (Tucker 2014).

In order to improve their users' sense of privacy, Facebook is one of the social networks that is investing in the development of technology to empower users to determine exactly what information they want to make available to the public and what data they prefer to withhold (Tucker 2014).

8.1.2   Business Issues

The successful deployment of social network sites in the business arena should follow specific guidelines. Moreover it is crucial to use measurable criteria to assess the actual effects of the use of online social networks in terms of revenue (Isaías et al. 2012a).

The participation on web based social networks has repercussions on the users' business relationships. It is believed that individuals with online social network presence have more opportunities to connect and strengthen ties with other professionals. Despite being hosted online, web based social networks are facilitating offline relationships (Benson et al. 2014). It is important that in the future, research approaches the connection between social network participation, professional communities' affiliations and the acquisition of workplace and career competences (Benson et al. 2014).

Social networks are excellent communication channels with unlimited audience reach and information dissemination. When examining the dynamic of event organisation, for example, it is possible to understand the important role that social networks play in event promotion. Organisers can use social networks as vehicles of information. In the case of music festivals, there is a significant amount of data that can be disseminated through social networks (performers, schedules, etc.) to those attending or wishing to attend the event. Additionally, the engagement of people in social networks is potentially beneficial in terms of building the attendees' loyalty to the event and again in terms of marketing the event with personal statements (text, photos, etc.) provided by the attendees (Hudson et al. 2015).

Within online social networks, there are members that work as "influential". These members can be very valuable for businesses due to their word-of-mouth power and their status of role models. They can reach their contacts more proficiently, which causes them to swiftly and widely disseminate information and by acting as role models, the other members are likely to be motivated to mimic them. The identification of these key users has become a central issue for business, so much so, that the strategies to make that identification have more recently become a significant research topic (Klein et al. 2015).

Online social networks have an important role in the empowerment of clients. They are interactive platforms that allow users to generate content, search for information and express their opinions about different products and brands. Internet users are sometimes called digital evangelists for their influential role among social networks, which can cause a product to proliferate or fail. Also, they are often denominated prosumers for their part in companies' creative process, via the suggestion of new products or services (Gonzalez et al. 2015).

8.1.3   Education Issues

Social networks can be used for formal or informal education (Teoh et al. 2014). The claims that online social network participation has a positive impact on students learning have been the focus of much research efforts (Thelwall and Kousha 2014; Park et al. 2014; Lawler and Molluzzo 2010; Vie 2008).

Social networks use in education remains a subject of interest due to their extensive reach, to the frequency and intensity with which they are used and their promising educational value (Park et al. 2014). Although there is research arguing that social networks mainly work as a distraction, there are also studies that attest to their positive role in enhancing communication and the relationship that students develop with teachers (Teoh et al. 2014) (Isaías et al. 2009). On the other hand, when social networks are used for intimidation or unwanted contacts, students can feel that these platforms are a mere extension of the challenges that they already face offline (Isaias et al. 2013a).

More recently, research is focusing on different aspects of social network use in education in order to potentiate its value. Teoh et al. (2014), for example, studied the role that gender plays in social network usage for learning. The authors concluded that male students are more prone to perceiving social networks as important pedagogical tools, than female students. This information, regardless of the limitations of the study, can be determinant for the implementation of social networks in classes with significant gender differences.

Social networks are being used in the education sector, but education is also being used in the online social networks arena. The increase of children and teenage users on social networks has lead to the preparation of several educational packages that promote a more secure participation on these platforms (Vanderhoven et al. 2014).

An important trend in the application of social networks in education is the creation of Mobile Social Network Sites (MSNS) for educational purposes. The iniquitousness use of social networks facilitated by mobile devices has created the notion of MSNS (Wang and Du 2014).

8.1.4   Other Sectors

The health sector has been focusing on the value that online social networks can represent for this area. Health entities' profiles on these platforms constitute an improvement in terms of their accessibility. Additionally, social networks can assist patients in the management of their pathologies. Their contribution to the collection of important data is also under scrutiny. On the one hand they are sources of unlimited and rich data, but on the other hand they pose reliability and bias challenges.

Moving forward, it is fundamental that research focuses on examining strategies to effectively use them as data sources (Alshaikh et al. 2014).

The health sector has an important preventative role, which relies greatly on the use of media to reach extensive audiences with health campaigns and awareness initiatives. Besides resorting to traditional media, the health sector has been investing in the use of internet media, namely the use of social networks. The recent interest in the use of online social networks to convey health information is based on their extensive reach, on the fact that the information can be transmitted to existing contacts, on their capacity for high engagement and retention levels, and on their interactive nature. The use of social networks to promote behavioural changes in terms of health is in a embryonic stage, but in the future research is expected to gain more insight into the actual benefits of these platforms for long term behaviour transformation (Maher et al. 2014).

Although web based social networks are being used for the purpose of social graphs analysis for quite some time, more recently, they are posing several challenges. Online social networks are growing in size, reach, complexity and data protection procedures. These changes are demanding advanced techniques for social graphs analysis. More specifically, this evolution of online social networks' characteristics have hindered sampling processes (Haralabopoulos and Anagnostopoulos 2013). The millions of users that compose online social networks pose a challenge in terms of its analysis as a whole. The search for a method that can produce a representative sample usually results into three types of graph sampling techniques: by random node selection, by random edge selection and by exploration. Nonetheless, these methods are incapable of creating a sample that can replicate the characteristics of an original graph (Yoon et al. 2015).

Haralabopoulos and Anagnostopoulos (2013) argue that different sampling techniques should be used in different situations to improve the identification of social network ties. In situations where the sampling size is small, Conventional Random Node Sampling should be used; in cases where larger samples are required, Enhanced Random Node Sampling is better suited. Yoon et al. (2015), on the hand, developed a sampling method that uses hierarchical community extraction and densification power law. By using these two techniques, the sampling method is able to generate sample graphs that reflect both the node-edge ratio and the topology of each region and of the entirety of the original graph. Additionally, subject recommended sampling techniques, such as snowball sampling, are also appropriate methods to assist the research of social networks (Isaías et al. 2013b).

Social networks, such as Facebook, have the potential to reach unlimited numbers of users, making them important resources for citizen participation and the promotion of causes and campaigns (Isaías et al. 2012b).

8.2   Web 3.0 Trends

This section addresses key Web 3.0 trends regarding Social Networks (SNs), organized by broad categories, as depicted in Fig. 8.2 below.

In Table 8.2, it can be summarized some of the Web 3.0 key trends detailed in the following sections.

Fig. 8.2 Social networks broad categories of Web 3.0 key trends (Prepared by Pedro Isaias)

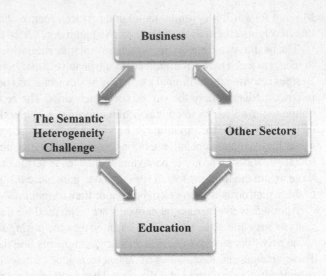

Table 8.2 Social networks web 3.0 key trends (Prepared by Pedro Isaias)

The semantic heterogeneity challenge	Business in SNs trends
Ontologies' definitions	Marketing Web 3.0
Data integration	Filtering possibilities
Semantic heterogeneity	E-commerce role
Semantic interoperability	Decision Support Systems (DSSs)
	Data integration
Education in SNs trends	**Other sectors in SNs trends**
Personalized learning objects	Social media and masses of information
Role in medical education	Ontologies and the tourism sector
Web 3.0 in MOOCs	Biomedical research
	e-Government
	Weather forecasting

8.2.1 The Semantic Heterogeneity Challenge

The overwhelming amount of information available online means that users can have access to unlimited sources of data, however that does not necessarily mean that data is more accessible. The volume of information available on the internet seems to be varying proportionally to the difficulty of extracting meaning from it. For this reason the Semantic Web aims to semantically interpret existing online data (Rana and Singh 2014). The fundamental concept of Web 3.0 is machine-understandable data. Hence, Web 3.0 has the challenging mission of adding meaning to online resources, through the definition of ontologies. This mission is particularly complex due to the fact that the Web has an open nature and as such "online semantics can be defined by different people, for different domains, and can vary significantly in expressiveness, richness, coverage, and quality, leading to increasing semantic heterogeneity." (Gracia and Mena 2012). When discussing the implications of semantic heterogeneity for the financial sector Li et al. (2014) refer to the creation of a "data *Tower of Babel*", which provides a clear illustration of the challenge of heterogeneity. Thus, semantic heterogeneity also poses a challenge for data integration (Jing 2015).

Both semantic ambiguity (different meanings for the same word) and semantic redundancy (different words for the same meaning) constitute an obstacle to the successful deployment of Semantic Web technologies. Semantic heterogeneity hinders the interoperability that is expected from Web 3.0 and despite the fact that this issue is addressed in specific domains and systems, there are only scarce solutions for dealing with it on a global scale (Gracia and Mena 2012).

In situations where applications are using competing ontologies their capacity to interoperate becomes compromised. Ontology matching is often used to address this issue (Shvaiko and Euzenat 2013), as it is regarded as one of the solutions to facilitate semantic interoperability. It consists in establishing a correspondence between similar semantic representations in ontologies (Rana and Singh 2014). There is a variety of matching systems such as SAMBO, Falcon, DSsim, RiMon, ASMOV, Anchor-Flood and AgreementMaker (Shvaiko and Euzenat 2013).

Maree and Belkhatir (2015) divide the different approaches to ontology alignment into three groups based on single-strategy, multiple-strategy and the exploitation of external semantic resources. The authors propose an alternative to these approaches by developing a framework that merges domain-specific ontologies using numerous external semantic assets (Maree and Belkhatir 2015).

As, an alternative to ontology matching Zadeh and Reformat (2013) presented a technique to identify semantic similarity that emphasises the relation between the terms and their semantics. This technique enables an evaluation of context-aware similarity and of specific segments of information that are part of the terms.

Data integration refers to the integration of data deriving from several sources and it can be used to solve semantic heterogeneity. Data integration has three approaches: data consolidation, data propagation and data federation. In addition, data integration can also use ontology to address heterogeneity. SIMS, OBSERVER, DOME, KRAFT, COIN are just some of the various systems that use ontology for this purpose (Sowmya Devi et al. 2014).

8.2.2   Business Issues

The concept of Enterprise 3.0 is becoming increasingly popular and it uses Web 3.0 as a platform (Ahrens and Zaščerinska 2014). Additionally, Marketing 3.0 has emerged, partly, due to changes in the behaviour of consumers. Clients have ambitions of a more collaborative and cultural marketing (Erragcha and Romdhane 2014).

This 3.0 version of marketing is based on the relationship between several actors, namely consumers, enterprises and sponsors. Clients have become increasingly creative and able to act as co-inventors of products. Additionally, globalization has made people value their culture, thus placing cultural matters in the list of priorities of commercial brands (Erragcha and Romdhane 2014).

With its constant evolution, the internet has supplied businesses with different type of information. In its primordial stage, Web 1.0, delivered information about products; Web 2.0, in its turn allowed insight into the customers' viewpoints; Web 3.0 uses all that information and transforms it into knowledge. The flow of information available hinders management decision making. Hence, Web 3.0 offers filtering possibilities and the opportunity and means to sort through the unlimited amounts of data. E-commerce is also an area where Web 3.0 can have a significant role. The semantic web can endow e-commerce businesses with features that will tailor the purchasing experience according to the clients' needs and characteristics, namely by using geo-referencing and client profile's data (Almeida et al. 2013).

Decision Support Systems (DDS) have been taking advantage of the features of Web 3.0 for the past decade. The Semantic Web can be applied to DDSs to assist several processes, namely, the integration and exchange of data, "web service annotation and discovery, and knowledge representation and reasoning." (Blomqvist 2014). Data integration is one of the major challenges of DDSs. By using Web 3.0 for data integration purposes, it is also possible to improve research, since more data is linked. Despite all of the advantages that Web 3.0 can represent in terms of DDSs, the scalability of the Semantic Web and its lack of maturity in optimisation and efficiency, which other more conventional methods of data management do offer (Blomqvist 2014).

8.2.3   Education Issues

While much debate exists still on the use of Web 2.0 in education, a more current discussion is the progressive use of Web 3.0 as an educational tool. The widespread use of the term e-Learning 3.0 is one of the indicators of Web 3.0's impact in education. The specific characteristics of Web 3.0 allow this version of the Web to afford nor only personalization, but also information management and semantic enrichment. The challenge for the future in terms of Web 3.0's deployment in education is the concrete steps that educators and students will take to incorporate it in their practices and routines (Miranda et al. 2014b).

Web 3.0 has become a resourceful enabler of personalized Learning Objects and Virtual Learning Environments (Kurilovas et al. 2014). It is, moreover, associated with the concepts of big data, cloud computing, augmented reality and 3D visualization, personal agents and with linked data (Dominic et al. 2014). The close relationship, between the Semantic Web and Artificial Intelligence, promises to endow the education sector with the capability to manage a great volume of data, since Artificial intelligence is a powerful tool for exporting meaning and patterns in data (Dominic et al. 2014). The unlimited sources and volumes of data available on the internet hinder its adequate use and application. The Semantic Web offers a solution by investing in sorting and categorizing information (Jiang 2014). Content is classified, structured, and endowed with specific annotations that enable its comprehension by machines. The use of ontologies attributes meaning to content and allows it to be exchanged and reused (Vera et al. 2013).

Medical education uses virtual patients to improve students' learning process, but their use across different systems can be very challenging. The authors developed a system (OpenLabyrinth) using the Semantic Web, that enables virtual patients' sharing and resource repurposing. This use of semantic annotation is becoming very important in repurposing content (Dafli et al. 2015).

The Semantic Web introduces new technologies and methods to link, edit and present information (Powell et al. 2012). Web 3.0 is also being used in MOOCs as a technological support for enhanced cooperation and communication (Waßmann et al. 2014). Furthermore, by the interaction and communication that takes place in learning environments can be used to tailor a more personalised learning experience (Halimi et al. 2014).

Moving forward in its role in education, Web 3.0 will have to address its interoperability challenges and also the issues deriving from ontology creation (Miranda et al. 2014a). Furthermore, Web 3.0 comes with additional security and privacy concerns (Dominic et al. 2014).

8.2.4   Other Sectors

According to Bontcheva and Rout (2014) "social media streams pose a number of new challenges, due to their large-scale, short, noisy, context-dependent, and dynamic nature." The colossal amount of information that is generated by social media can no longer be addressed by conventional search approaches. The Semantic Web is being regarded as an alternative to the conventional methods in the sense that it can assist user to manage the overload of data that originates from social media. The use of automatic semantic-based methods can be beneficial for both data interpretation and decision making processes of media streams, by being able to adjust to the users' information search objectives (Bontcheva and Rout 2014).

The use of ontologies within the tourism sector has the potential to minimize the detrimental effect of different depictions of tourist locations. By using ontologies it is possible to create a structured foundation of common depictions (Nikola et al. 2014). Semantic destination management systems offer complete integration, flexibility, and personalization. They have the ability to combine marketing and management into multiple products and services; they offer the flexibility to integrate single tourist destinations; and the targeted information that they supply, the services can be personalized to meet the customers' needs (Nikola et al. 2014).

The amount of data that biomedical research involves presents a challenge in terms of its analysis. The data is abundant in quantity, in types of format and in sources, which hinders data integration and interoperability. Translational medicine works towards minimizing the cleavage between research and medical practice. The accomplishment of its mission statement relies greatly on data integration and interoperability, thus, translational medicine has been focusing on Web 3.0 for its capacity of semantic depiction and data interoperability. The systems that are already in use that implement Web 3.0 technologies have proven efficient in terms of public and private data integration, semantic representation, and knowledge extraction. The challenge for the future is to help these semantic web systems to evolve from a local-scale approach to a network of collaboration and partnership (Machado et al. 2015).

The e-Government sector is responsible for numerous services that involve both their national borders and their international relations, and a variety of agencies. This mission implies the management of colossal amounts of data deriving from a multiplicity of sources, which is hindered by insufficient automation and interoperability. In order to address these challenges, semantic web technologies can be considered. Liu et al. (2013) suggest applying semantic business process management to e-Government by designing a framework that consists in four layers: data, process, semantic and presentation. This framework uses semantic technologies allied with business process management to improve automation, interoperability and data integration and reuse. Additionally, it is important to invest in methodologies for ontology development. In order to take advantage of Semantic Web technologies, it is imperative to develop a government domain ontology (Dombeu and Huisman 2011).

Weather forecasting information is central to a panoply of sectors and as the number of different systems, formats and parameters become involved in producing information, more strategies need to be put in place to assure its quality. The use of semantic technologies in this field is essential to facilitate the integration of data form multiple sources and the interoperability between different applications and systems. The employment of the semantic web maximizes the potential of knowledge integration in an area where the accuracy of the information is determinant (Ramar and Mirnalinee 2014).

8.3 Conclusion and the Future

The above depicts the need for novel interfaces capable of coping with issues of (i) variety, (ii) dimensionality and (iii) scalability. Moreover, these novel interfaces must be able to deal with multi-media information, 3D and augmented realities, and be able to adjust to various sectors and *niche* markets. There is a world of developments to evolve in the near future and the reader is invited to seat at the front row of these developments.

References

Ahrens A, Zaščerinska J (2014) Analysis of teachers' use of web technologies: focus on teachers' enterprise 3.0 application. J Inf Technol Appl Educ 3(1):25–32

Almeida F, Santos JD, Monteiro JA (2013) E-commerce business models in the context of web3. 0 paradigm. Int J Adv Inf Technol (IJAIT) 3(6):1–12

Alshaikh F, Ramzan F, Rawaf S, Majeed A (2014) Social network sites as a mode to collect health data: a systematic review. J Med Internet Res 16(7):e171

Benson V, Morgan S, Filippaios F (2014) Social career management: social media and employ-ability skills gap. Comput Hum Behav 30:519–525

Blomqvist E (2014) The use of semantic web technologies for decision support–a survey. Semantic Web 5(3):177–201

Bontcheva K, Rout D (2014) Making sense of social media streams through semantics: a survey. Semantic Web 5(5):373–403

Dafli E, Antoniou P, Ioannidis L, Dombros N, Topps D, Bamidis PD (2015) Virtual patients on the semantic web: a proof-of-application study. J Med Internet Res 17(1):e16

Dominic M, Francis S, Pilomenraj A (2014) E-learning in web 3.0. Int J Mod Educ Comput Sci 6(2):8–14

Erragcha N, Romdhane R (2014) New faces of marketing in the era of the web: from marketing 1.0 to marketing 3.0. J Res Mark 2(2):137–142

Dombeu JVF, Huisman M (2011) Combining ontology development methodologies and semantic web platforms for e-government domain ontology development. Int J Web Semant Technol (IJWesT) 2(2):12–25

Frost J, Vermeulen IE, Beekers N (2014) Anonymity versus privacy: selective information sharing in online cancer communities. J Med Internet Res 16(5):e126

Gonzalez R, Llopis J, Gasco J (2015) Social networks in cultural industries. J Bus Res 68(4):823–828

Gracia J, Mena E (2012) Semantic heterogeneity issues on the web. Internet Comput, IEEE 16(5):60–67

Halimi K, Seridi-Bouchelaghem H, Faron-Zucker C (2014) An enhanced personal learning envi-ronment using social semantic web technologies. Interact Learn Environ 22(2):165–187

Haralabopoulos G, Anagnostopoulos I (2013) Real time enhanced random sampling of online social networks. J Netw Comput Appl 41:126–134. doi:http://dx.doi.org/10.1016/j.jnca.2013.10.016

Hudson S, Roth MS, Madden TJ, Hudson R (2015) The effects of social media on emotions, brand relationship quality, and word of mouth: an empirical study of music festival attendees. Tour Manage 47:68–76

Isaías P, Miranda P, Pífano S (2009) Designing e-learning 2.0 courses: recommendations and guidelines. In: Méndez-Vilas A, Solano Martín A, Mesa González JA, Mesa González J (eds) Research, reflections and innovations in integrating ICT in education, vol 2. FORMATEX, Badajoz, pp 1081–1085

Isaias P, Pífano S, Miranda P (2012a) Social network sites: modeling the new business-customer relationship. In: Safar M, Mahdi KA (eds) Social networking and community behavior model-ing: qualitative and quantitative measures. IGI Global, Hershey, pp 248–265

Isaías P, Pífano S, Miranda P (2012b) Web 2.0: harnessing democracy's potential. In: Ed D, Matthew AJ (eds) Public service, governance and web 2.0 technologies: future trends in social media. IGI Global, Hershey, pp 223–236

Isaias P, Miranda P, Pífano S (2013a) The impact of web 2.0 adoption in higher education. In: Kommers P, Kasparova E, Bessis N (eds) Proceedings of IADIS international conference web based communities and social media. Prague, Czech Republic, pp 65–73

Isaias P, Pífano S, Miranda P (2013b) Subject recommended samples: snowball sampling. In: Isaias P, Nunes MB (eds) Information systems research and exploring social artifacts: approaches and methodologies: approaches and methodologies. IGI, Hershey, pp 43–57

Jiang D (2014) What will Web 3.0 bring to education? World J Educ Technol 6(2):126–131

Jing H (2015) Analysis of the problem of semantic heterogeneity in the integration of railway system. Int J Hybrid Inf Technol 8(1):427–434

Kataoka H, Ogawa Y, Echizen I, Kuboyama T, Yoshiura H (2014) Effects of external information on anonymity and role of transparency with example of social network de-anonymisation. In: Availability, Reliability and Security (ARES), 2014 9th international conference on, 8–12 Sept 2014, pp 461–467. doi:10.1109/ARES.2014.70

Klein A, Ahlf H, Sharma V (2015) Social activity and structural centrality in online social networks. Telematics Inform 32(2):321–332

Kurilovas E, Kubilinskiene S, Dagiene V (2014) Web 3.0–Based personalisation of learning objects in virtual learning environments. Comput Hum Behav 30:654–662

Lawler JP, Molluzzo JC (2010) A study of the perceptions of students on privacy and security on social networking sites (SNS) on the internet. J Inf Syst Appl Res 3(12):1–18

Li T, Lemieux VL, Pottinger R (2014) Challenges in resolving semantic heterogeneity with the global legal entity identifier system. In: Proceedings of the international workshop on data science for macro-modeling, 2014, ACM, pp 1–2

Liu Z, Le Calvé A, Cretton F, Evéquoz F, Mugellini E (2013) A framework for semantic business process management in e-government. In: Proceedings of the IADIS international conference WWW/INTERNET 2013, pp 259–267

Machado CM, Rebholz-Schuhmann D, Freitas AT, Couto FM (2015) The semantic web in translational medicine: current applications and future directions. Brief Bioinform 16(1):89–103

Maher CA, Lewis LK, Ferrar K, Marshall S, De Bourdeaudhuij I, Vandelanotte C (2014) Are health behavior change interventions that use online social networks effective? A systematic review. J Med Internet Res 16(2)

Maree M, Belkhatir M (2015) Addressing semantic heterogeneity through multiple knowledge base assisted merging of domain-specific ontologies. Knowl-Based Syst 73(0):199–211. doi:http://dx.doi.org/10.1016/j.knosys.2014.10.001

Miranda P, Isaias P, Costa CJ (2014a) From information systems to e-learning 3.0 systems's critical success factors: A framework proposal. In: Zaphiris P, Ioannou A (eds) Learning and collaboration technologies. Designing and developing novel learning experiences. Springer International Publishing, Cham, pp 180–191

Miranda P, Isaias P, Costa C (2014b) The impact of web 3.0 technologies in e-learning: emergence of e-learning 3.0. In: Proceedings of the EDULEARN14, pp 4139–4149

Nikola M, Angelina N, Jelena TC (2014) The impact of web 3.0 technologies on tourism information systems. Paper presented at the SINTEZA, Serbia

Park SY, Cha SB, Lim K, Jung SH (2014) The relationship between university student learning outcomes and participation in social network services, social acceptance and attitude towards school life. Br J Educ Technol 45(1):97–111

Peddinti ST, Ross KW, Cappos J (2014) On the internet, nobody knows you're a dog: a twitter case study of anonymity in social networks. In: Proceedings of the second edition of the ACM conference on Online social networks, Stanford University, California, ACM, pp 83–94

Powell M, Davies T, Taylor KC (2012) ICT for or against development? An introduction to the ongoing case of Web 3.0. IKM Emergent Research Programme, European Association of Development Research and Training Institutes (EADI)

Ramar K, Mirnalinee TT (2014) A semantic web for weather forecasting systems. In: Recent trends in information technology (ICRTIT), 2014 international conference on, 10–12 Apr 2014, pp 1–6. doi:10.1109/ICRTIT.2014.6996127

Rana V, Singh G (2014) An analysis of semantic heterogeneity issues and their countermeasures prevailing in semantic web. In: Optimization, reliabilty, and information technology (ICROIT), 2014 international conference on, 6–8 Feb 2014, pp 80–85, doi:10.1109/ICROIT.2014.6798296

Shvaiko P, Euzenat J (2013) Ontology matching: state of the art and future challenges. IEEE Trans Knowl Data Eng 25(1):158–176

Sowmya Devi L, Barathi J, Hema M, Chandramathi S (2014) A survey: different approaches to integrate data using. Ontol Methodol Improve the Qual Data 2(11):126–131

Teoh K-K, Pourshafie T, Balakrishnan VA (2014) Gender lens perspective of the use of social network in higher education in Malaysia and Australia. In: Proceedings of the 2014 international conference on social computing, Beijing, ACM, p 21

Thelwall M, Kousha K (2014) Academia. edu: social network or academic network? J Assoc Inf Sci Technol 65(4):721–731

Tucker CE (2014) Social networks, personalized advertising, and privacy controls. J Mark Res 51(5):546–562

Vanderhoven E, Schellens T, Valcke M (2014) Educational packages about the risks on social network sites: state of the art. Procedia-Soc Behav Sci 112:603–612

Vera MMS, Breis JTF, Serrano JL, Sánchez M, Espinosa PP (2013) Practical experiences for the development of educational systems in the semantic web. NAER: J New Approaches Educ Res 2(1):23–31

Vie S (2008) Digital divide 2.0: "Generation M" and online social networking sites in the composition classroom. Comput Composition 25(1):9–23. doi:10.1016/j.compcom.2007.09.004

Wang R-B, Du C-T (2014) Mobile social network sites as innovative pedagogical tools: factors and mechanism affecting students' continuance intention on use. J Comput Educ 1(4):353–370

Waßmann I, Schönfeldt C, Tavangarian D (2014) Wiki-Learnia: social E-learning in a web 3.0 environment. Eng Sci Technol/Nauki Inzynierskie i Technologie 4(1):21–27

Yoon S-H, Kim K-N, Hong J, Kim S-W, Park S (2015) A community-based sampling method using DPL for online social networks. Inf Sci 306(0):53–69. doi:http://dx.doi.org/10.1016/j.ins.2015.02.014

Zadeh P, Reformat MZ (2013) Assessment of semantic similarity of concepts defined in ontology. Inf Sci 250(0):21–39. doi:http://dx.doi.org/10.1016/j.ins.2013.06.056

Index

Printed in the United States
By Bookmasters